UNDENIABLE

Evolution and the Science of Creation

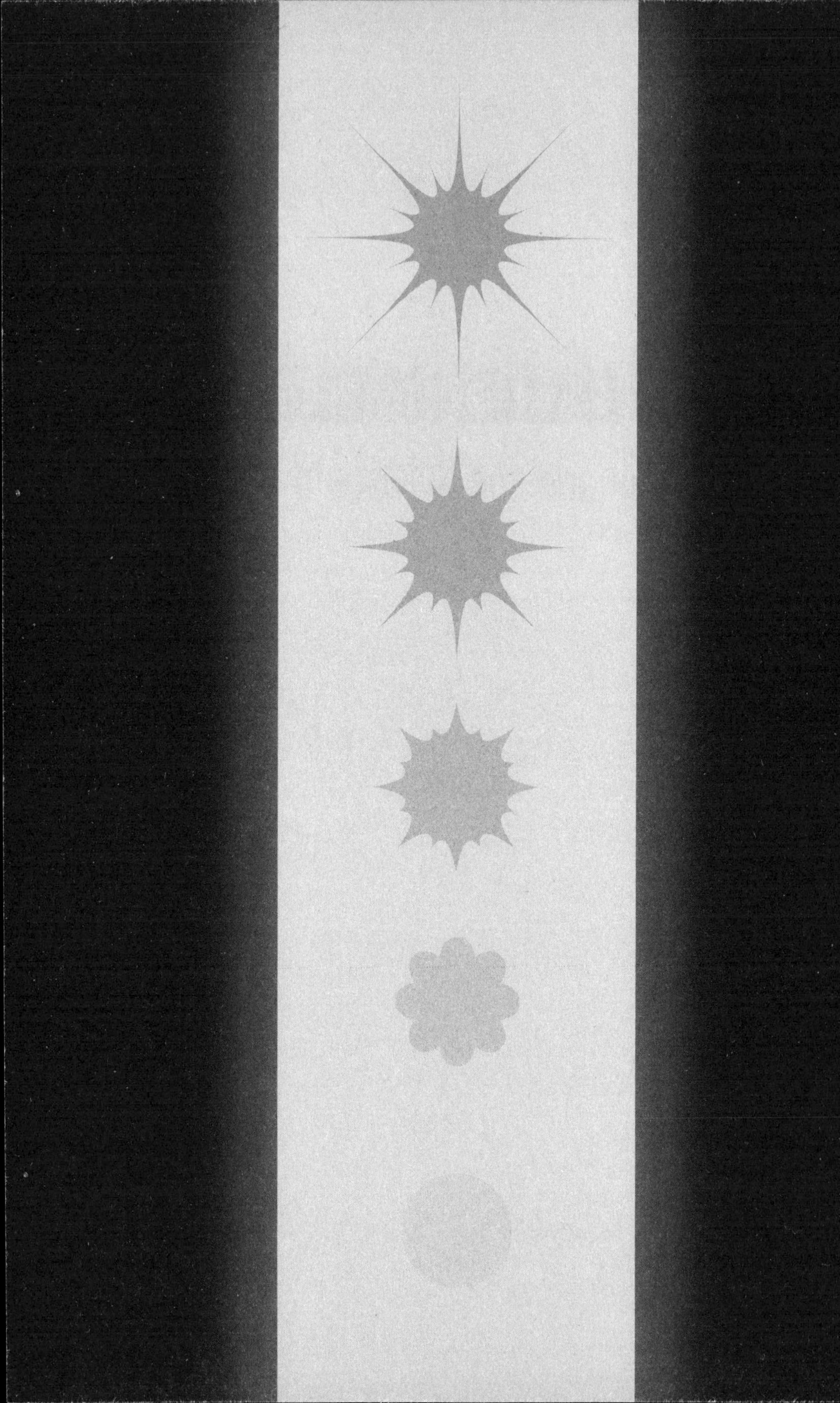

UNDENIABLE

Evolution and the Science of Creation

by

BILL NYE

edited by

COREY S. POWELL

ST. MARTIN'S PRESS NEW YORK

www.stmartins.com

Design by Omar Chapa

Library of Congress Cataloging-in-Publication Data

Nye, Bill.

Undeniable: evolution and the science of creation / Bill Nye with Corey Powell.—First edition.

pages cm

ISBN 978-1-250-00713-1 (hardcover)

ISBN 978-1-4668-6988-2 (e-book)

1. Natural history—Philosophy—Popular works. 2. Evolution (Biology)—Popular works. 3. Creationism—Popular works. 4. Religion and science. I. Powell, Corey S., 1966– II. Title. III. Title: Evolution and the science of creation.

QH45.5.N94 2014

576.8—dc23 2014027163

St. Martin's Press books may be purchased for educational, business, or promotional use. For information on bulk purchases, please contact Macmillan Corporate and Premium Sales Department at 1-800-221-7945, extension 5442, or write specialmarkets@macmillan.com.

First Edition: November 2014

10 9 8 7 6 5 4 3 2 1

Dedicated to science students of all ages,
wishing you countless safe journeys
and the infinite joy of discovery

CONTENTS

UNDENIABLE

Evolution and the Science of Creation

1

ME AND YOU AND EVOLUTION, TOO

I think it started with the bees. I was about seven years old, and I watched them . . . all day. That Sunday, I had read the "Ripley's Believe It or Not" column in *The Washington Post*, which claimed, "The Bumblebee: Considering its size, shape, and wingspan, is an aerodynamic misfit—which should be unable to fly!" It was frustrating, because here they were flying. I got caught up in the details. Their wings looked like decoration, no more useful than a store-bought bow glued to a gift. I looked closely at my mother's azalea flowers—so many delicate parts. Somehow, the bees were able to get in there, fill their pollen baskets from the flowers, and fly away again and again.

How did bees learn to do all that? Where did they come from? Where did the flowers come from? Come to think of it, how did any of us get here? Why did Ripley's have it so obviously wrong? I was getting pulled into something much larger than myself. The yearning to know about nature and where or how we fit in is deep within all of us. As I learned about evolution and descent by natural selection, the answers fell into place.

We are all aware that evolution happens, because we all have parents. Many of us have, or will have, children. We see the effects of heredity up close and personal. We've also experienced firsthand what Charles Darwin called descent with modification: the way that an entire population of living things can change from generation to generation. Think about the food grown on farms. For about twelve thousand years, exploiting the phenomenon of evolution, humans have been able to modify plants through a process known as artificial selection. In wheat farming and horse racing we call it breeding. Darwin realized that breeding (and domesticating) plants and animals involves exactly the same process that occurs naturally in evolution, only accelerated with the help of humans. This natural process produced you and me.

Once you become aware—once you see how evolution works—so many familiar aspects of the world take on new significance. The affectionate nuzzling of a dog, the annoying bite of a mosquito, the annual flu shot: All are direct consequences of evolution. As you read this book, I hope you will also come away with a deeper appreciation for the universe and our place within it. We are the results of billions of years of cosmic events that led to the cozy, habitable planet we live on.

We experience evolution every day in our culture as well. People everywhere are fascinated with other people. That's why we have sidewalk cafés, televisions, and gossip magazines. We interact to produce more of us for future generations. People are fascinated with their bodies. Turn on the television to any channel. If it's youth-oriented music programming, you'll see advertisements for skin medicines to make you look healthy, for deodorants to modify your natural scents, and for hair and makeup products to render you more attractive to a potential mate. If it's a staid news channel, you'll see ads for improving your breathing, your bones, and, of course, your sexual performance.

None of these products would be produced were we not walking, talking products of evolution.

We are all so much alike, because we are all human. But it goes deeper than that. Every species you'll encounter on Earth is, near as we can tell, chemically the same inside. We are all descended from a common ancestor. We are shaped by the same forces and factors that influence every other living thing, and yet we emerged as something unique. Among the estimated 16 million species on Earth, we alone have the ability to comprehend the process that brought us here. Any way you reckon it, evolution is inspiring.

Despite all of that, a great number of people in many parts of the world—even in well-educated parts of the developed world—are resistant or hostile to the idea of evolution. Even in places like Pennsylvania and Kentucky, here in the United States, the whole idea of evolution is overwhelming, confusing, frightening, and even threatening to many individuals. I can understand why. It's an enormous process, unfolding over times that dwarf a human lifespan—across billions of years and in every part of the world. And it's profoundly humbling. As I learned more about evolution, I realized that from nature's point of view, you and I ain't such a big deal. Humans are just another species on this planet trying to make a go of it, trying to pass our genes into the future, just like chrysanthemums, muskrats, sea jellies, poison ivy . . . and bumblebees.

Many people who are troubled by evolution want to suppress teaching the whole concept of descent through natural selection in schools. Others try to push it aside or dilute it by casting doubt on the established science that supports it. State education standards allow the teaching of fictitious alternatives to evolution in Texas, Louisiana, and Tennessee. Even though the people who support these curricula live lives that are enriched in many ways by science and engineering (everything from running water and abundant food to television and the Internet) they avoid the exploration of evolution, because it reminds us all that humankind may not be that special in nature's scheme. What happens to other species also happens to us.

I continually remind people what is at stake here. Our understanding of evolution came to us by exactly the same method of scientific discovery that led to printing presses, polio vaccines, and smartphones. Just as mass and motion are fundamental ideas in physics, and the movement of tectonic plates is a fundamental idea in geology, evolution is *the* fundamental idea in all of life science. Evolution has essential practical applications in agriculture, environmental protection, medicine, and public health. What would the deniers have us do? Ignore all the scientific discoveries that make our technologically driven world possible, things like the ability to rotate crops, pump water, generate electricity, and broadcast baseball?

Even the theological objections to evolution stand on shaky ground. For the last century and a half, ever since the publication of Darwin's *On the Origin of Species* in 1859, many people have come to believe that evolution is in conflict with their religious beliefs. At the same time, many people around the world who hold deep religious convictions see no conflict between their spiritual beliefs and their scientific understanding of evolution. So the naysayers are not only casting doubt on science and nonbelievers; they are also ignoring the billions of non-conflicted believers around the world, dismissing their views as unworthy.

I'll admit that the discovery of evolution is humbling, but it is also empowering. It transforms our relationship to the life around us. Instead of being outsiders watching the natural world go by, we are insiders. We are part of the process; we are the exquisite result of billions of years of natural research and development.

Frankly, my concern is not so much for the deniers of evolution as it is for their kids. We cannot address the problems facing humankind today without science—both the body of scientific knowledge and, more important, the process. Science is the way in which we know nature and our place within it.

Like any useful scientific theory, evolution enables us to make predictions about what we observe in nature. Since it was developed in the nineteenth century, the theory itself has also evolved, by which I mean that it's been refined and expanded. Some of the most wonderful aspects and consequences of evolution have been discovered only recently. This is in stark contrast to creationism, which offers a static view of the world, one that cannot be challenged or tested with reason. And because it cannot make predictions, it cannot lead to new discoveries, new medicines, or new ways to feed all of us.

Evolutionary theory takes us into the future. As the foundation of biology, evolution informs big questions about emerging agricultural and medical technologies. Should we genetically modify more of our foods? Should we pursue cloning and genetic engineering to improve human health? There is no way to make sense of these issues outside of an evolutionary context. As an engineer trained in the U.S., I look at the assault on evolution—which is actually an assault on science overall—as much more than an intellectual issue; for me, it's personal. I feel strongly that we need the young people of today to become the scientists and the engineers of tomorrow so that my native United States continues to be a world leader in discovery and innovation. If we suppress science in this country, we are headed for trouble.

Evolutionary theory also takes us into the past, offering a compelling case study of the collaborative and cumulative way that great scientific discoveries are made. In some sense the concept of evolution can be traced to the Greek philosopher Anaximander. In the sixth century BC, after evaluating fossils, he speculated that life had begun with fishlike animals living in the ocean. He had no theory of how one species gave rise to another, however, nor did he have an explanation of how Earth acquired its stunning biodiversity. Nobody would, for another two millennia. Ultimately, the mechanism of evolution was discovered by two men at very nearly the same time: Charles Darwin and Alfred Wallace.

You've probably heard a great deal about Darwin. You may not have heard so much about Wallace. He was a naturalist who spent a great deal of time in the field studying and collecting specimens of flora and fauna. He traveled in the Amazon River basin and in what is now Malaysia. Through his far-flung geographic and intellectual explorations, Wallace formulated his theory of evolution independent of Darwin, and described an important aspect of the evolutionary process, often still referred to as the "Wallace effect" (more about that in chapter 12). Wallace recognized humans as just one part of a much broader living world. Quoting from his 1869 book *The Malay Archipelago*, ". . . trees and fruits, no less than the varied productions of the animal kingdom, do not appear to be organized with the exclusive reference to the use and convenience of man . . ." In Victorian England, such a point of view was controversial to say the least.

Darwin had the earlier start. Wallace was just eight years old in 1831 when the twenty-two-year-old Darwin had a remarkable opportunity as an energetic young man to go to sea aboard the HMS *Beagle*. He realized that if humans could turn wolves into dogs, then new species could come into existence by the same means naturally. He also saw that populations do not grow and grow indefinitely, because their environment will always have limits on the resources available. Darwin connected these ideas by observing that living things produce more offspring than can survive. The individuals compete for resources in their respective ecosystems, and the individuals that are born or sprout with favorable variations have a better shot at survival than their siblings. He realized that, left unchecked, the process of natural selection would result in the great diversity of living things that he would go on to observe.

Recognizing the two scientists' convergent views, colleagues arranged for Wallace and Darwin to present a paper together at a meeting of the Linnaean Society in London in 1858. The paper was based on a letter that Wallace had written to Darwin, along with an abstract for a paper that Darwin had written in 1842. The revolutionary impact of the joint presentation was not immediately obvious to all of those in attendance. Thomas Bell, the president of the Linnaean Society, infamously reported that no important scientific breakthroughs had occurred that year: "The year which has passed has not, indeed, been marked by any of those striking discoveries which at once revolutionize, so to speak, the department of science on which they bear . . ."

The publication of *On the Origin of Species* in 1859 created a sensation and proved President Bell spectacularly wrong. It also made Darwin far more famous than Wallace, as Darwin remains to this day. His ability to articulate the theory of evolution is still astonishing. *On the Origin of Species* remains a remarkable and remarkably readable book, readily available in hardback, paperback, and online a century and a half later. In it, Darwin gives us example after example of evolution and explains the means by which it happens, providing both the facts and the mechanism in one volume.

Evolution is one of the most powerful and important ideas ever developed in the history of science. It describes all of life on Earth. It describes any system in which things compete with each other for resources, whether those things are microbes in your body, trees in a rain forest, or even software programs in a computer. It is also the most reasonable creation story that humans have ever found. When religions disagree about just creation, there is nothing to do but argue. When two scientists disagree about evolution, they confer with colleagues, develop theories, collect evidence, and arrive at a more complete understanding. Every question leads to new answers, new discoveries, and new smarter questions. The science of evolution is as expansive as nature itself.

Evolution goes a long way toward answering the universal question that ran through my brain as a kid, and still does: "Where did we come from?" It also leads right into the companion question we all ask: "Are we alone in the universe?" Today, astronomers are finding planets rotating around distant stars, planets that might have the right conditions for supporting life. Our robots are prospecting on Mars looking for signs of water and life. We're planning a mission to study the ocean of Jupiter's moon Europa, where there is twice as much seawater as there is here on Earth. When we go seeking life elsewhere, the whole idea of what to look for, and where to look for it, will be guided by our understanding of evolution. Such a discovery would be profound. Proving that there is life on another world would surely change this one.

The great questions of evolution bring out the best in us: our boundless curiosity, and our boundless ability to explore. After all, evolution made us who we are.

2

THE GREAT CREATIONISM DEBATE

For those readers who might be deeply religious, welcome. I very much hope you make it through this chapter. It's about my recent debate with a creationist in the Commonwealth of Kentucky, which in many ways was the impetus for me to write this book. Our issue was whether or not creationism is "viable" (the term agreed upon) as an explanation of . . . well, of anything. I emphasize that I did not disparage anyone's religion. I did not mention anything about *The Bible.* I had no reference to Jesus from the city of Nazareth. But I was, and remain, concerned about the extraordinary claim that Earth is extraordinarily young, which is an assault not just on evolution but on the whole public understanding of science.

Having a few thousand people make use of a few million dollars to promote their point of view is not unusual. This is actually what a great many not-for-profit organizations do, including the Union of Concerned Scientists, the National Center for Science Education, and my own Planetary Society. It's also part of how government policies are developed and put into law. In the case of creationism however, certain not-for-profit groups set out to indoctrinate our science students in their central idea: that the first book of *The Bible*'s assertion that Earth is only six or ten thousand years old (the exact number depends on their interpretation) is supported by scientific evidence. Such an idea is laughable and could be easily dismissed were it not for the political influence of these groups. In general, creationist groups do not accept evolution as the fact of life. It's not just that they don't understand how evolution led to the ancient dinosaurs, for example, they take it another step and deny that evolution happened at all anywhere, let alone that it is happening today. They want everyone else in the world to deny it, too, including you and me.

Inherent in this rejection of evolution is the idea that your curiosity about the world is misplaced and your common sense is wrong. This attack on reason is an attack on all of us. Children who accept this ludicrous perspective will find themselves opposed to progress. They will become society's burdens rather than its producers, a prospect that I find very troubling. Not only that, these kids will never feel the joy of discovery that science brings. They will have to suppress the basic human curiosity that leads to asking questions, exploring the world around them, and making discoveries. They will miss out on countless exciting adventures. We're robbing them of basic knowledge about their world and the joy that comes with it. It breaks my heart.

I got the chance to write this book after expressing my concern about the future of the U.S. economy on an Internet Web site called BigThink.com. I pointed out that without young people entering science fields, especially engineering, the country will fall behind other nations who do educate their kids in real science rather than the pseudoscience of creationism. Subsequent to that, I was challenged to a debate by Ken Ham, an Australian-born evangelical leader who has managed to oversee the construction of an amazing building that he calls the Creation Museum in Kentucky. His organization is called Answers in Genesis. He claims that his interpretation of *The Bible* is more valid than the basic facts of geology, astronomy, biology, physics, chemistry, mathematics, and especially evolution.

After a few months of mulling it over, I agreed to go to the Creation Museum and take the guy on head-to-head, or lectern-to-lectern. I chose to participate in this debate to raise awareness of the creationist movement and its inherently deleterious effects on our society, as it dulls our resolve to tackle big scientific challenges like producing energy for the burgeoning human population. Perhaps it's not surprising that along with his other extraordinary claims, my opponent doesn't feel that he or his followers should be concerned with climate change.

We were each given time to make our case before the audience. Mr. Ham holds to a fascinating pair of doublespeak phrases: "observational science" and "historical science." He says that there's a difference between things that happen while you're alive and watching and things that happened before you were born. So for him, anything in the fossil record is subject to question. For him, any astronomical observation is automatically irrelevant, because the stars are older than any person that could have observed them. Perhaps a mischievous deity put them all there in a flash. Using the word *science* in these Orwellian ways is unsettling. As a science educator, I also find this practice deeply irresponsible.

When it was my turn, I hammered away at Mr. Ham's claim that there was a big ole flood and that all the animals we see today are descendants of the few pairs that Noah and his family were able to save on a big boat, the ark of Biblical myth. By the way, neither *The Bible* nor Mr. Ham offers any insight into the fate of every surface-dwelling plant during this supposed episode.

I started by discussing stratigraphy, the layering (strata) of the rocks that make up Earth's crust. I could not help but point out that the Creation Museum building sits atop millions of years of limestone layers. The famous landmark Mammoth Cave is right there in Kentucky not far away. On the way to the event, I had no trouble finding a piece of limestone with a small shelly ancient sea creature fossil clearly visible. It was just off the shoulder of Interstate 69. I showed the audience photos of the Grand Canyon, including the striking Muav limestone, the Temple Butte formation, and the Red Wall limestone. They are 505 million, 385 million, and 340 million years old, respectively. What's so striking about them is how distinct they are from one another. Clearly there were three different spans of time during which these three very-different-colored deposits were formed.

I made the significant paleontological point that in each of these strata, there exist certain fossils that are specific to that time. Impressions of creatures like trilobites that lived in the more distant past are found in the lowermost layers. Creatures like ancient mammals that lived in the most recent times are found in the topmost layers. And creatures that lived in between are found in the in-between layers. There is no place, not one single example, of a fossil from one layer trying to swim its way up into a more recent deposit. If there were a great flood and every living thing was drowning all at once, we would expect one of them somewhere to be caught trying to save its life. There is not one instance of this in any stratum anywhere on Earth. If you find one, you will turn science on its head. You will be famous. Believe me, people are looking.

Early in my presentation, I talked about the ice cores, long cylinders of ice extracted by researchers from ice sheets (in Greenland and Antarctica, especially). There are samples with 680,000 layers of snow-ice. Every year a layer of snow falls. It gets compacted by each subsequent year's precipitation. I asked how there could be 680,000 layers, if there weren't 680,000 seasons of snow (in other words a period of 680,000 years). I explained that in Ham's natural history, you'd have to have 170 winter-summer seasons every single year, for every time Earth went around the Sun. Such a turn of events is just not possible.

Did you know that there are bristlecone pine trees in the western United States that are significantly older than 6,000 years? If you put a tree under water for a year, you kill it, and that's exactly what would have happened according to the creationists' literal reading of Noah's flood. There is a tree in Sweden called Old Tjikko that is apparently 9,550 years old. I, and apparently most of the online audience, was thinking: For cryin' out loud Mr. Ham, what sort of weird world do you live in? If a tree is 9,000 years old, the Earth is not 6,000, etc.

As a lover of math, this was a fun one: Mr. Ham claims there were 7,000 kinds of animals on Noah's ark; there are about 16 million species extant today (that's my very conservative estimate based on recent surveys of life). To get from 7,000 species 4,000 years ago to 16 million today, we'd need to find 11 new species every day. Not every year! And not 11 individual animals! Eleven new species would need to be identified every single day! It's a multiplication and division problem. Not difficult, but very difficult to refute.

It was fun for me also to point out that this brand of young-Earth creationism claims that kangaroos came from a huge ship, the ark, which is supposed to have safely run aground on Mount Ararat in modern-day Turkey. It's a respectable peak—5,165 meters (almost 17,000 feet)—and it's snowcapped. It's not clear to me how all the animals and humans made the arduous descent. The kangaroos, both of them, are supposed to have made it down the mountain, ran or hopped from there to Australia—and no one saw them. Furthermore, if they took a reasonable amount of time to make the trip, you'd expect some kangaroo pups or joeys to have been born and some adults to have died along the way. You'd expect some kangaroo fossils out there somewhere in what is now Laos or Tibet. Also, they are supposed to have run across a land bridge from Eurasia to Australia. But there's no evidence of such a bridge or any kangaroo fossils in that area, not any.

Speaking of the ark itself, I pointed out that very skilled shipwrights in New England built the *Wyoming*, a six-masted wooden tall ship. The boat is huge by wooden ship standards, over 100 meters (300 feet) long. The creationists' imagined ark is said to have been 500 feet long and able to hold 14,000 animals and 8 people. The real ship, the *Wyoming*, had a crew of 14. Although it was built by the best shipbuilders in the world in 1909, they could not manage the inherent elasticity of wooden timbers and strakes. The *Wyoming* twisted in rough seas, opening uncontrollable leaks in the hull. She foundered and sank, with the loss of all hands. If the best in the modern world couldn't build a large seaworthy ship, what reason would anyone have to think that 8 unskilled ancient people could?

I pointed out the National Zoo in Washington, D.C., has about 400 species on 66 hectares (163 acres) of land. Zookeepers work around the clock to maintain the health of those wonderful creatures. How could 8 unskilled people keep 14,000 animals alive and well? Doesn't seem to me that they could.

I pointed out the spectacular boulders one can see along highways in the Pacific Northwest region of the U.S. They were washed there by ancient floods, when ice dams periodically gave way in what is now Montana. If there had been a worldwide flood, and the heavier rocks sank to the bottom as Mr. Ham et al. assert, what are these boulders doing there on top of the ground and not under the soil? Well, they wouldn't be there, but they are. So, the creationists are wrong about the natural history of their world.

I mentioned also the key feature of any scientific theory, whether it's about evolution or anything else: One should be able to use it to make predictions. I briefly mentioned the remarkable story of *Tiktaalik*, the "fishapod" (transition between fish and tetrapod, or land animal with four legs), whose fossil was predicted to exist in certain types of swamps from the Devonian Period. Scientists led by the tireless Neil Shubin of the University of Chicago found exactly such a fossil swamp in northeastern Canada, and went there and found *Tiktaalik*. Think about it, an ancient animal was surmised to have once existed. Researchers figured out where it would have lived. They went there and proved it. Amazing.

I also used the reproductive strategies of modern Mexican topminnows as another example of a scientific theory being used to make a prediction. Topminnows reproduce asexually when they have to. In those episodes, their offspring are more susceptible to infection by the blackspot parasitic worm, because they have a less varied mix of genes. That sexual strategy is exactly what is predicted by adjunct theories of evolution, notably one called the Theory of the Red Queen, which just charms me. In the fictional land of the Red Queen, Lewis Carroll's Alice has to run all the time. Evolution is thought to work the same way: If you stop running, stop mixing up your genes, you'll fall off the treadmill of life. The queen will leave you behind. I've devoted a whole chapter to this idea farther in.

Because I personally met astronomer and Nobel laureate Robert Wilson, I very much enjoyed reminding the audience of his discovery, in conjunction with Arno Penzias, of cosmic microwave background radiation. In the 1960s, the two of them found that the whole sky is glowing, which is exactly what cosmologists who worked on the theory of the Big Bang had predicted. I asked also how we could observe stars that are farther away than 6,000 light-years, if Earth is only 6,000 years old. One would expect to see no light at all from such places, unless natural laws are overthrown for a while. So why do we see far more distant stars and galaxies in all directions? If there were a superpower, why would it (she or he) mess with us that way?

For his part, Ken Ham avoided responding to any of these issues and repeated that he had "a book," and his interpretation of said book supersedes anything we can observe in nature. I pointed out that his interpretation of his book is just not reasonable to anyone who examines the world with an open mind and curiosity. This foreshortened worldview is not at all consistent with the views of many of the world's religious leaders. Finally, to tie it to my main concern, I reminded the audience of Article 1, Section 8 of the United States Constitution. One of the duties of Congress is "to promote the progress of Science and the Useful Arts . . ."

One of the duties of all of us—parents, scientists, everyone—is to help educate the next generation, so they can succeed in their lives and help make the world a better place. Watching some of Mr. Ham's videos in preparation for the event, I could not help but notice his kvetching about young people, specifically that they're leaving his ministry. After hearing him out for more than an hour of his monologues, I gathered that they have trouble taking what he says to heart.

The creationists press on, looking for ways to isolate their kids as much as possible and to indoctrinate them so thoroughly that no matter what the world throws at them, the children will grow up to do their best to accept a 6,000-year-old Earth. All the while, the whole lot of them have no issue embracing modern information technology, medicine, and food systems that enable them to conduct their extraordinary business. As I often remark, it wouldn't matter, if it weren't for the kids.

The essence of the evening was captured by a question from the audience. Someone asked: "What would it take to change your worldview?" My answer was simple: Any single piece of evidence. If we found a fossilized animal trying to swim between the layers of rock in the

Grand Canyon, if we found a process by which a new huge fraction of a radioactive material's neutrons could become protons in some heretofore fantastically short period of time, if we found a way to create eleven species a day, if there were some way for starlight to get here without going the speed of light, that would force me and every other scientist to look at the world in a new way. However, no such contradictory evidence has ever been found—not any, not ever.

Mr. Ham responded that *nothing* would change his mind. He has a book that he believes provides all the answers to any natural science question that could ever be posed. No piece of evidence would change his mind—not any, not ever. Imagine this man or some of his followers on a jury. If their minds were made up, there would be nothing for the defense or prosecuting attorneys to do. No evidence would sway these jurors. They would refuse to use their intellect to assess the quality of evidence. They would not employ even the most rudimentary critical thinking skill. They would sit very politely, I imagine, but evidence would not matter at all. That is a very troubling prospect indeed. The rule of law would be ignored. They would be, well they are, excluding themselves from our society. They do not want to participate. I hope all of us will consider the potential consequences of this sort of thinking—or nonthinking. If there were a test of competency for voters, how well would they fare?

At some level, as an altruistic human (a consequence of my evolutionary heritage), I feel bad for the creationists. They have been left out of the wonderful process of science and its ability to reveal so much about nature. I'm heartbroken for their kids. On top of that, I feel bad for all of us. How did we let an ideological resistance to inquiry become such a prominent part of our society? How did we exclude so many people from the knowledge won with great sacrifice by our ancestors? Perhaps in the coming decades we can turn this around and include everyone—people in nonscientific fields and professions as well as the professional scientist, engineer, and educator. Perhaps by celebrating evolution, we can open minds and unlock more of our vast human potential.

Despite the resistance from part of my audience (not to mention from a large portion of the American public), I suspect everyone there can reason along the lines I described in my half-hour presentation. Certainly I suspect that Ken Ham can. It's just that when it comes to evolution, and especially to the related realization that we are all pretty small bits of the universe, it seems as though Ham and his followers just can't handle the truth. They throw aside their common sense and cling to the hope that there's something that makes it okay to *not* think for themselves. The irony is, in the process they are walking away from our ability to understand who we are, where we came from, and how we fit into a cosmos of astounding dimensions. If there is something divine in our nature, something that sets humans apart from all other creatures, surely our ability to reason is a key part of it.

Ken Ham, his followers, myself, and everyone else—we are all in this together. We're all a product of the same evolutionary processes. Here's hoping we can work together to bring the children of the creationists' preachers' flocks to a more enlightened, boundless way of thinking about the world around us.

3

CREATIONISM AND THE SECOND LAW OF THERMODYNAMICS

It's not just Ken Ham and his Answers in Genesis ministry. Over the years, I've heard a lot of arguments against evolution from people who find it objectionable on religious or emotional or philosophical grounds. Often these disputes boil down to the simple, dead-end argument from incredulity: "It cannot be true, because I find it so hard to believe that it's true." But sometimes creationists take a more interesting, science-inspired line of attack and insist that evolution is not physically possible, because no system can naturally become more complex over time. More specifically, they cite evolution for violating one of the most well-established principles in science, the Second Law of Thermodynamics.

In commonsense terms, the Second Law is this: Given the chance, balls roll downhill; they never roll uphill on their own. Put another way, energy tends to spread out: Heat spreads out, and lakes never spontaneously freeze on a warm summer day. Creationists seem to think that the human species has likewise been running downhill since The Fall, since our ancestors Adam and Eve screwed up. Creationists hear the Second Law of Thermodynamics and say, "Ah, hah! See, our whole world is a machine winding down—death to everyone."

By the way, rest assured that there is a First and Third Law of Thermodynamics; there's even a Zeroth Law. While these are cool (sorry) in their own way, they don't come up in creationists' diatribes.

To be sure, the Second Law of Thermodynamics really does contribute to a general winding down of the world around us. It explains why no one can build a perpetual motion machine. Somewhere, someplace in any machine, you're going to lose some energy to heat. When it comes to making something go or happen, there's no free lunch. The following quotation is irresistible; it's from the renowned twentieth-century astronomer Arthur Stanley Eddington:

"The law that entropy always increases—the Second Law of Thermodynamics—holds, I think, the supreme position among the Laws of Nature. If someone points out to you that your pet theory of the universe is in disagreement with Maxwell's equations—then so much the worse for Maxwell's equations. If it is found to be contradicted by observation—well, these experimentalists bungle things sometimes. But if your theory is found to be against the Second Law of Thermodynamics I can give you no hope; there is nothing for it but to collapse in deepest humiliation."

You don't have to know or care about Maxwell's equations to get a sense of what he means. (But just so you know, they are the equations that describe the nature of light, electricity, and magnetism.) The key idea is that the Second Law of Thermodynamics mathematically describes any system's loss of energy to its surroundings. It is fundamental to the way the natural world works. Since energy is constantly and continuously spreading out, everything should be winding down to a dead stop. Perhaps you can see why creationists figure that the Second Law means there's no way for evolution to add complexity to life. How could any living system get organized, if all its driving forces were continually diluted, spread out into the vast blackness of the universe?

As a mechanical engineer who took a lot of physics, I am fascinated by this particular creationist argument, because it is both scientifically subtle and completely misinformed. Here's the most important thing to know: The Second Law applies only to closed systems, like a cylinder in a car engine, and Earth is not even remotely a closed system. Transfers of matter and energy are constantly taking place. Life here is nothing like a perpetual motion machine, but neither is it like a ball rolling inexorably downhill.

There are three main sources of energy for life on Earth: the Sun, the heat from fissioning atoms deep inside Earth, and the primordial spin of Earth itself. These sources provide energy throughout the day. The Sun provides the most energy. It's a fusion reactor releasing 10^{26} Watts every second (10^{26} Joules). Earth's core also provides energy in the form of heat. The spinning of our planet keeps shifting the energy inputs and adds acceleration to the wind and the waves. So as you can see, the world we live on is not even remotely a closed system. All of our world's ecosystems ultimately run on a continual external source of light and heat. Energy has been pouring in from the Sun for over four and a half billion years. Living things ranging from amoebas to sequoias have to find ways to make the best use of all that energy, lest they be outcompeted by other living things that use it more efficiently.

The Second Law sets the boundaries; it's the rule we all have to play by. Starting with energy to study evolution is a great way to understand life. What do living things do with all this energy? We use it to drive chemical systems that obey the Second Law. But the Second Law comes into play everywhere in your life. When you pedal a bicycle, there's a little bit of friction in the chain joints and the bearings that hold the pedal and cranks. The motion makes a little bit of heat. Where does that heat go? It goes into the universe. Really. It dissipates into the environment of the whole world and eventually radiates into space, and there's no way to recover it. The tendency for energy to spread out in natural systems might also explain how a kid's room becomes such a mess.

The modern mechanical world runs on heat, and it's also bounded by the Second Law. Car engines, jet airplane turbines, and coal-fired power plants all use heat produced by burning to make something spin. The heat comes from chemical reactions. The same is true for us animals. Instead of fire in our bellies, produced by combining carbon fuel with oxygen, we have enzymes that enable chemicals in our food to combine with oxygen to produce chemical energy. But whether it's a turboprop or a cricket, we all must produce a bit more energy than we are able to use. We lose some fraction of our energy to the universe; the Second Law constrains us.

From time to time, people pursue this bit of logic: If heat is always spreading out, won't the entire cosmos cool down to some fantastically cold state, with nothing moving at all, anywhere? Will there be a heat death of the universe (sometimes known as the "Big Freeze")? If such a state is destined to be, it will occur unimaginably far into the future. The universe is 13.8 billion years old, and from our vantage point, it seems to be just getting under way.

Now back to those creationists who go all crazy insisting that since disorder always increases, and heat spreads out, Earth and everything on it must be getting more and more disordered. They are completely wrong that the Second Law puts a lid on complexity, because of their confusion (or deception?) about closed versus open systems. But it turns out that the analysis of the flow of energy, especially with the Second Law in mind, is a wonderful way to approach evolution. It provides a useful way to understand the way that living things use the energy that's available. I hope you read every word of my book, but there's more on this particular idea in the next chapter as well as in chapters 29 and 35. Read on.

Evolution is also not random; it's the opposite of random. One of Darwin's most important insights is that natural selection is a means by which small changes can add complexity to an organism. With each generation of offspring, the beneficial modifications can be retained. Each mutation that doesn't work as well in nature either dies off with the organism directly, or gets outcompeted by others of its kind in succeeding generations of offspring. It's by the process of evolution that beneficial changes are added up and up and up.

The fundamental energy needed for these beneficial mutations to evolve comes primarily from the Sun and the molten insides of Earth. Decaying organic matter consumed by living beings provides the chemical energy for each beneficial mutation. And each successive generation can carry all of the beneficial mutations. The mechanism for adding complexity is in having offspring. Each generation leading up to any given organism's existence used energy from the Sun to provide its nutrition and warmth.

Far from violating the Second Law of Thermodynamics, evolution is a powerful validation of the law. Embracing the idea that life adheres to the Second Law is akin to saying that evolution is not random; living things are directed and selected by competition. Life runs on energy, and likewise evolution runs on energy. A system that makes use of energy almost serves as a definition of life. There's an intriguing new twist on thermodynamic energy and life; we mention it again in chapter 35. Earth's whole ecosystem comprises organisms expending energy collectively, competing, and developing new forms. That is what makes life so marvelous: It channels energy into butterflies and *Arabidopsis* (first plant to have its genome sequenced) and sea jellies and people.

That's what makes the creationist viewpoint not just staggeringly wrong, but sadly impoverished. In twisting around the Second Law of Thermodynamics, they take a powerful tool for understanding the world and try to make it into a barrier to understanding instead. But there is a silver lining here. By inspiring people to learn the fundamental features of nature described by the Second Law of Thermodynamics, creationists can actually inspire a richer appreciation of the mechanism of evolution.

4

BOTTOM-UP DESIGN

When I was a senior in college, earning a degree in mechanical engineering, I was recruited by the Boeing Commercial Airplane Company to work on the 747. Rest assured that I was very well supervised. Like most human organizations, Boeing is organized from the top down. The company was started by Bill Boeing himself; he hired the people he wanted and assigned them to desks and drawing boards, organizing his business in top-down fashion. To this day, Boeing has a top-down structure headed by a CEO, a president, a board of directors, and a chairman of the board. It's an arrangement familiar to anyone who has worked for, or with, a large corporation. It's also a big reason for a lot of popular misconceptions about the process of evolution.

In top-down organizations, everything follows a chain of command. At work you might have an organization chart (an "org" chart) that shows this chain: the boss at the top, a layer of managers beneath her or him, a layer below those middle managers, including shop foremen, team leaders, and entry-level employees. The same pattern plays out in all other kinds of hierarchical groups. If you're a student, you can pretty much count on there being a principal or president and a vice or assistant principal at the school. Universities are loaded with presidents, deans, department chairs, ombudsmen, professors, and teaching assistants.

Nature follows an organizational scheme, too, but one that's stunningly different from ours—and that's where the confusion can creep in. Humans like to organize things from the top down; many of us reasonably assume everything is organized that way. But nature works the other way around. In the natural scheme of things, changes made in the past are the only things that determine whether or not any feature of the organization is retained in the future. There's no planning. If there were a day-to-day manager of nature, he or she would have a cushy gig, because he or she wouldn't have to do anything. Nature is self-organizing. That's another way of defining evolution: Nature builds ecosystems, in all of their complex glory, from the bottom up.

Looking at nature with a human's top-down perspective can create a mistaken sense of intentional design. I'll show you what I mean. Let's say you've started a business, and your organization is successful enough that you are able to hire a few people. Your business grows, and as it does, it gets more complicated and requires more energy. More computers are needed. More phones are needed. You need more equipment and more energy to supply everything from copy machines to farm irrigators. All this equipment and all these people have to be organized. The more complicated it gets, the more organizing it needs. The energy enabling this organization and this growth comes from outside of the company. If you sell things or services, your business growth comes from your environment: in this case, from the money spent by your customers.

In nature, living things depend on their environment, as well. We get energy stored in chemical bonds in our food; plants generally get their energy from sunlight; a few ecosystems run on geothermal or volcanic heat. When we view our systems and nature's systems from the standpoint of energy, our organizations and nature have a lot in common. However, there is a big difference between the two. Any decisions you make to shape and direct your business are based on what resources are available, but they are your decisions. You directed your organization to make certain purchases, hire certain people, and fill out certain paperwork, or whatever documentation you might require. Your company or business gets more complex, because you chose to make it so.

In nature, living things also have the ability to use the resources in their environments to become more complex, not by their conscious choices, but by outcompeting other living things. This is one of the fundamental mechanisms of Darwinian evolution: natural selection. The chemicals along a strand of DNA are arranged so that the molecule can make a copy of itself. Things being the way they are out there (or way down in there), these copies are not perfect. In the same way, you can tell the difference between an original document and a copy of the document made with a copy machine, it's very difficult in nature to make a perfect copy. Those small changes in DNA that occur during an organism's developmental stage result in the organism being just slightly different from its parents or parent organism. They introduce variation within a population. These changes can help an organism live and eventually reproduce, hinder the organism's reproduction, or produce no noticeable difference. You can see why people might think these changes result from conscious or willful acts, but they don't.

The changes that help an organism reproduce stay with the organism's offspring; beneficial changes get passed on in the DNA. When the offspring, in turn, reproduce, they have that beneficial characteristic, and that helps them produce offspring themselves later on. The changes that hinder an organism keep some fraction of that organism's population from reproducing. So hindering changes don't get passed on; they disappear from future versions of that organism's DNA. The changes that don't make any difference—don't make any difference. They get passed on as well.

In general, when energy becomes available to an organism, that energy helps an organism survive and reproduce. Incoming energy (from food or sunlight) can drive and create a beneficial change, which can lead to increased complexity in the offspring of those living things. Once Charles Darwin saw this connection, he realized what a powerful idea it is.

You can contrast the system in nature with a system in a human-built organization like a corporation. Hardly anything happens to benefit an organization unless someone somewhere makes a choice. Very few changes happen organically, that is, from the organization being affected automatically. Someone has to step in and hire or fire, invest or divest, buy or sell, or else nothing happens. Certainly nothing happens automatically to make a system more complex in a good way. You might say that human organizations depend on an intelligent designer.

As a corporation grows, different divisions add systems, paperwork, forms to fill out, hoops to jump through, and so on, to help their division get things done. At some point a manager might come in and analyze that the organization is top-heavy, too many middle managers managing too few people below them. He or she might determine that there is too much paperwork, too much redundant storage of transactions or records, and so on. Then that manager starts cutting pieces off or trying to streamline things.

It doesn't work that way in evolution. If you have a system that holds an organism back and keeps it from reproducing with success, that organism will not pass its genes to the next generation. Nobody has to decide anything. Although a change in a gene usually happens at random, the next generation of that gene is subject to forces that are anything but random. You've got the right combination of genes or you don't. You're still in the game, or you're not. We call it selection pressure; it determines which genes get through.

Many creationists and science deniers, especially in the United States, cite randomness as part of the process of evolution and go on to insist that since evolution is random it cannot explain the rich complexity of life. This is essentially another form of the thermodynamics argument I talked about in the previous chapter. Creationists often use the example of a hypothetical tornado swirling its way through a hypothetical junkyard whose contents include all the pieces to build one of my beloved old 747s. (This is sometimes called the junkyard tornado argument.) What are the chances, they ask, that you'd end up with a perfectly assembled, operable airplane? Obviously, zero, because it would be random.

The problem with this argument is that the premise is wrong. Evolution, and the selection of reproduction-worthy genes that drives it, is the opposite of random. It is a sieve that living things have to pass through successfully, or we never see them again. At Boeing—well, at any company—there are selection pressures that work quite a bit like natural selection. There is competition between airplane companies. Customers and airline corporations that buy billions of dollars' worth of planes want their equipment to be efficient. They want their planes to use less fuel, to be easy to maintain, and to be cheaper overall, because those things are expensive. So, managers, engineers, machinists, interior designers, ergonomic experts—everybody works to make the planes faster, better, and cheaper.

When I was in engineering school, my aeronautics professor showed us that winglets were gimmicks, a waste of time and energy. (You've seen them. Winglets are the little vertical pieces on the tips of modern airliner wings.) Airplanes and birds are able to fly because the air pressure under the wing is higher than the air pressure above it. It's a result generally of tipping the wing up in the front a little, giving it what's called an angle of attack. It works for 787s and barn owls. The higher pressure under the wing induces air to squirt around the wing tip. As the plane or owl moves through the air, it leaves a continuous whirlpool behind it. Spinning up the atmosphere in this fashion takes energy. It robs the plane or bird of a little efficiency. Winglets block a great deal of the tip spinning, and so improve energy efficiency, but they also add weight. My old professor had us do this analysis assuming aluminum wings and aluminum winglets. What we didn't account for, at least the first time through, was the invention of lighter weight, strong plastic composite materials.

Today, planes have composite plastic winglets. It's a form of evolutionary selection pressure. It's a result of market forces, but it's still human-caused decision-making. A company that did not embrace that technology might end up selling fewer planes and going out of business. Winglets are a result of countless hours of research and development. They result from management decisions, engineering analysis, and fabricators' skill.

Here's the amazing thing: Barn owls have a style of winglets, too. There is no evidence that they were deliberately designed that way. Instead, owl winglets are the product of generation after generation of owls reproducing and occasionally producing babies (owlets) with feathers that suppressed wing tip vortices just a little better than others of their species. That trait got passed on and on and on, without any org chart.

It's not too hard to imagine a corporation, given countless years, with the ability to fire every employee that was not as good as another one. Eventually, after millions of employees came and went, the company would be the best in the world at doing whatever it did.

Boeing actually has employed over a million different people since it was founded almost a century ago, but Boeing has not had nearly as much time to work things out as nature has. Nevertheless, we see both bottom-up and top-down processes at work. Airplane designs have been tried and discarded, just like bottom-up evolution, and the end solution looks (not surprisingly) similar to the one that emerged from evolution. But the airplane designs were created *de novo* (from new) by human brains in a distinctly human, top-down organization.

We are at once constrained to play the hand that creation deals us, and empowered to come up with our own top-down methods to create our world the way we want it to be. We can employ our evolution-given brains to fly in planes and use our imagination to soar in spirit. We are a result of evolution, and therefore so are our creations—both the not-so-good and very good. It's glorious.

5

A DEEP DIVE INTO DEEP TIME

The idea that just leaving the world alone for a really, really, really long time can lead to all the different kinds of life we see seems incredible, at least until you appreciate the enormous timescale of evolution. Since the late eighteenth century, scientists have used the term "deep time" to describe the magnitude of the scales involved. Understanding just how deep the deep past really is has been likened to staring into an abyss. It's too deep to see the bottom, too deep to imagine. It can overwhelm your thoughts. But once you embrace such depths, the mechanisms of evolution begin to make sense.

The events that led from the first living cell to you and me have required a nearly unimaginable period of time. When we're talking about evolution, the expression "a long time" is an understatement. For me, here's a case in which it is an understatement to even use the expression "an understatement." Earth is currently reckoned to be 4.54 billion years old. Based on fossilized mats or layers of bacteria, we figure life got started here at least 3.5 billion years ago.

These dates have been determined through extraordinary insight and diligence by astronomers, biologists, chemists, geologists, and geochemists. I remember very well sitting in a meeting with my beloved senior colleague Bruce Murray, who exerted great influence over the whole American planetary exploration program. At this meeting, I remarked that a certain researcher in Europe was a geologist and should have insights into some of the business we were discussing. Bruce slapped his open hand on the tabletop demanding attention. He yelled at me, "That man is no geologist! He's a geochemist!" Wow, *excuse me*, Bruce. I must have grown up in some remote illiterate part of the world, where we do or did not make the distinction.

As usual, though, Bruce made a good point. Geochemists do a lot of the most important work in reckoning the age of ancient rocks . . . and if we don't appreciate what they do, we are missing out on a vital part of Earth's story. It's just a little over a century since the French physicist Henri Becquerel discovered radioactivity, and with it the key to unlocking deep time. Since then, physicists have developed extraordinarily successful models of the behavior of atoms. Atoms are made of protons, neutrons, and electrons. Protons and neutrons are, in turn, made of quarks. Energy can come and go, carried by photons and neutrinos, and so on. By studying certain elements carefully, we have observed that, for example, radioactive Rubidium-87, containing 37 protons (and 50 neutrons), can change or decay to strontium, which has 38 protons. These two elements can be thought of as a radiochemical system.

When rocks are liquid or nearly liquid, what geologists call plastic, they contain a certain amount of rubidium and a certain amount of strontium. The mixture is measurable by diligent radiochemists. When that molten rock spews out of a volcano, say, it solidifies. By looking at the ratios of certain elements frozen in with rubidium and strontium, radiochemists and geochemists can determine how long ago the melt, as it's called, turned solid. In the case of rubidium and strontium, we can count on precisely half of the rubidium-87 to transmute to strontium-87 (now with 49 protons) in 48.8 billion years. That's right, almost 50 billion (with a b) years. It is the nature of radioactivity. You cannot determine what any one atom will do, but you can determine with just crazy precision how long it will take a sample of half the stuff to change from one to the other. This is where the expression half-life comes from. Furthermore, we can determine when a quarter of it will change, when an eighth of it will change, a sixteenth, a thirty-second, a sixty-forth, a one-hundred-twenty-eighth, a two-hundred-fifty-sixth, etc.

The word *chemistry* is the key in this business of geochemistry and radiochemistry. Rocks in Earth's crust typically contain a certain amount of rubidium and a certain amount of strontium, along with other elements like calcium and potassium. The chemical behavior of rubidium is a lot like the chemical behavior of potassium, and the chemical behavior of strontium is a lot like the chemical behavior of calcium. (In chemistry we see that they reside in the same columns on the periodic table of the elements.) When the rocks are liquid, rubidium tends to remain free and unattached, but as the rock cools, rubidium sometimes takes the place of potassium in the rock crystals. In the same way, strontium substitutes for calcium. So by closely examining crystals that we know contain potassium and comparing the relative abundances of rubidium and strontium that are also in the crystal, we can determine the age of the rocks relative to other rocks. We can work our way into the past, pulling our date reckoning back in time by our bootstraps.

There are several other geochemical clocks that radiochemists use to reckon the age of Earth, besides rubidium-strontium. There's uranium-lead, there's potassium-argon, and there's samarium-neodymium. Each clock reckons time using different chemical elements and each provides us with incontrovertible evidence of Earth's age. You may have heard of carbon dating or carbon-14 dating. That is a related technique that is well suited to measuring shorter timescales. It lets us work backward in time to determine when a living thing stopped transpiring (plants) or stopped breathing (animals). Carbon dating only goes back a few tens of thousands of years, because the half-life of this type of carbon is only 5,730 years. Compare that with rubidium-strontium; this radiochemical clock goes back into the past almost a million times further. Carbon dating is important for studying human history, but it's not well suited to reckoning deep time.

Evolution snaps into focus when you realize how fantastically old our planet is. To imagine it, try this. Look at a map of North America. To readers from other parts of the world, it's interesting to note that what we often call the continental U.S. extends from the Atlantic to the Pacific oceans (as do Canada and Mexico). By means of the U.S. Interstate Highway System, one can drive a car from coast to coast. If we were to go from around San Diego on the Southwestern shore of the U.S. to, let's say, Boston on the Northeastern shore, we would go about 4,500 kilometers or 2,800 miles. That much driving would take you from Lisbon, Portugal to Moscow, Russia. And along the way, you would have passed through eight different countries.

Imagine a time line running from coast to coast. Let's say that for every kilometer of travel (every ten football fields for U.S. readers), you pass through one million years of time. Well then, every meter (about every yard) represents one thousand years of time. For this charming thought-model, the distance from your chin to your outstretched arm represents a thousand years. A Thousand Years!

For the next step, imagine yourself walking through time from San Diego to Boston. When you start, Earth is still a big orange-hot ball of molten rock. After two hundred kilometers, six or seven days into your hike, look up as you come across a marker reminding you that the Moon is forming. Another two days of walking, and a marker tells you that enough rain has fallen on a surface cool enough to have the ocean form; that was 4.4 billion years ago. After a month on foot, you'll be coming upon the first signs of life, about 3.5 billion years ago. Two thousand kilometers from your embarkation point, somewhere near Broken Arrow, Oklahoma, you'll find tiny microbes, blue-green bacteria. Before that, by the way, you would have suffocated along your route, because there was no significant amount of oxygen in the air. Earth's oxygen was a byproduct of photosynthesis in those early microbes. The blue-green bacteria, along with you and me, are the only single species known to be capable of altering the climate of an entire planet.

Two months into your trek, perhaps not too far from Little Rock, Arkansas, the ancient supercontinent of Rodinia has formed. Another month on the road, and you may notice that a more recent ancient supercontinent, Pangaea, has formed. Not all, but almost all of the living things you encounter are in the ocean. Wait, you wouldn't encounter them, unless you were walking when most of the current interior of the United States was under water, under an ancient inland sea. As you slog forward, unusual and by our standards bizarre sea creatures abound.

When you're only 230 kilometers (120 miles) from the eastern shore, you finally come upon the ancient dinosaurs. They are latecomers in the long history of life. You walk in their midst for 100 kilometers. That would be two or three days at a good pace. Along the way, plants that produce flowers appear. Sex is everywhere now.

Just two kilometers to go now, and the Atlantic Ocean might be in view. And here, you meet the first of us—early versions of humans, living just 2 million years ago. Keep on; you might meet some of our cave-dwelling ancestors. Within five meters of the water's edge, the ancient pyramids appear. Now, within twenty centimeters, not even the distance from your pinky fingertip to the end of your thumb, the United States as a nation comes to be. The human landing on the Moon is just two centimeters, less than an inch, from the water. Press your toes forward and you arrive at today.

Now turn around. Look back across the vastness of the continent. Most of it looked barren or desolate on your trek. All that we know of history, all the people and their affairs, everything you've come to know, takes place in less than your last stride. It's this vastness of time that has enabled life to begin and evolution to direct the creation of all the living things we've ever known.

Notice that during about three quarters of your hike, living things were just revving up. There were bacteria, lots of them. But the plants that you and I eat, along with the animals we raise for food and fertilizer, all came to be when you had almost completed your journey. Most of life's time here on Earth has been spent making the slow evolution from a few crudely self-copying chemicals to the first true cells to relatively uncomplicated, but nevertheless remarkable, living things. It is only very recently in the deep timescale of things that complicated animals like you, me, and my bewitching old girlfriend came to be.

When Charles Darwin and Alfred Wallace were pondering the consequences of their discoveries, they were deeply troubled by what seemed to be the tremendous amount of time required to get life to where it is now, or where it was when they lived. Darwin published *On the Origin of Species* in 1859. Radioactivity wasn't even discovered until 1896, and wasn't well understood until many years after that. So even as Darwin developed his elegant theory, which he established through dozens of remarkable, diligently executed clever experiments, he was constrained by a lack of a reasonable explanation for how evolution could have enough time to act. He couldn't explain how Earth could be so fantastically old.

Darwin's contemporaries challenged him, even ridiculed him, for asserting that all the living things we've ever seen on Earth have a common ancestor and came into existence over this vast expanse of time. How could it be? How could that much time have passed? It is still unimaginable for most of us, let alone to people in Darwin's day.

Through the late-nineteenth century, William Thompson (who was Irish but went by the uniquely British sobriquet of Lord Kelvin) provided the science world with a seemingly authoritative calculation that Earth was between 20 million and 400 million years old, tops.

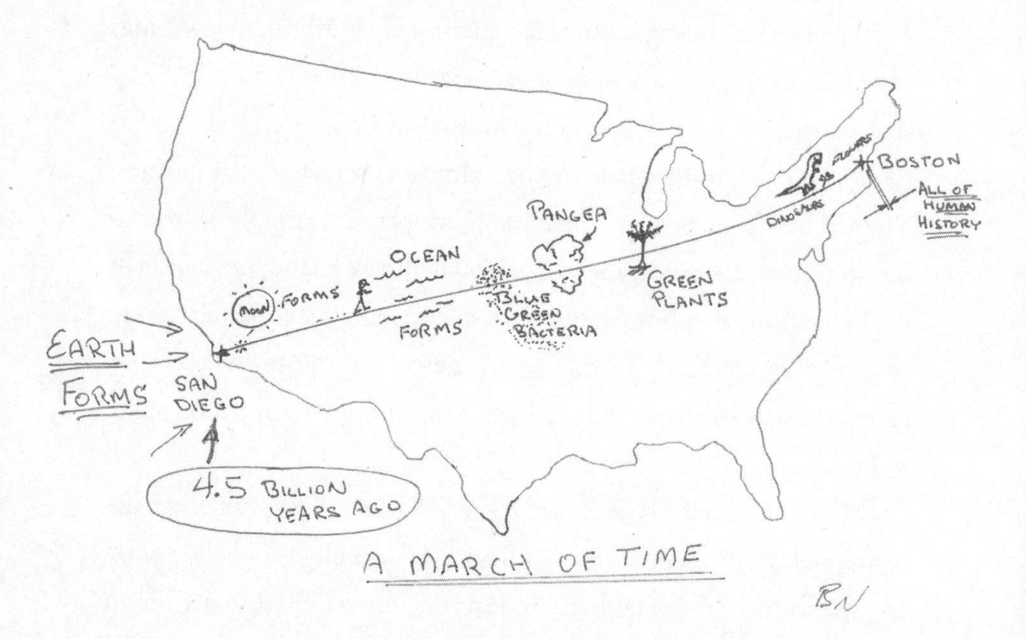

Evolution seemed to require ten or a hundred times as much history. It was a paradox. The true age of Earth remained a mystery through Wallace's and Darwin's lives. It wasn't until the discovery of radioactivity that scientists grasped the answer. Kelvin had assumed that Earth had been cooling ever since its birth, and he used its present temperature to deduce its age. What he didn't know is that radioactive elements deep inside our planet keep adding new heat. His calculations were perfect, but his understanding was not. In fact, there was plenty of time for evolution to unfold—every bit as much as Darwin imagined, and then some.

I often reflect on what an extraordinary time (pun intended) it is to be alive here in the beginning of the twenty-first century. It took life billions of years to get to this point. It took humans thousands of years to piece together a meaningful understanding of our cosmos, our planet, and ourselves. Think how fortunate we are to know this much. But think also of all that's yet to be discovered. Here's hoping the deep answers to the deep questions—from the nature of consciousness to the origin of life—will be found in not too much more time.

6

ON THE ORIGIN OF EVOLUTION

While my family was seated together eating a chicken dinner, sometime back in the 1960s, my father described a scene around his family's table that I like to think of as his version of Darwin's trip to the Galapagos: the moment when he realized that all living things are related. My grandparents lived in a pretty big house in Washington, D.C. To supplement their income during the Great Depression, my grandmother rented rooms to young men, students, or people just starting out in their careers. One of these guys would often sit at the dinner table and remark offhandedly about the close similarities between a chicken's knees and our knees, along with other unsettling anatomical parallels.

According to legend, my grandmother was okay with these observations, and her approval carried with it an acceptance of evolution and natural selection. She was quite the naturalist and spent a great deal of time studying wildflowers. But my grandfather, who attended church regularly, was troubled by all this. A link between humans and chickens flew in the face (or beak) of his churchgoing upbringing. The tenant was very nice and paid his rent on time, but his chicken talk was not-so-subtly influencing the kids—my father, his brother, and their friends. These dinners gave my father food for thought for the rest of his life.

My dad returned from World War II and went to work as a salesman. Nevertheless, he often referred to himself as Ned Nye, Boy Scientist. My mom spent that war as a lieutenant in the Navy. She was recruited because she was good at math and science; she went on to earn a doctorate in education. So I was brought up with a great deal of respect for the human capacity to figure things out and solve problems. As a kid growing up in Washington, I also had access to the Smithsonian Institution. I was often dumped off . . . er, I mean, I was often given a ride to the downtown bus and encouraged to visit the museums and see the sights. Like any kid, I was utterly fascinated by the ancient dinosaurs. I thought about how cool it would be to see one of those animals in the wild. Evolution was in my bones, you might say. I was immersed in the scientific story of life on Earth pretty much from the time my own life began. It's no wonder I ended up writing this book.

Now consider what things were like when Charles Darwin and Alfred Wallace were formulating their ideas in the first half of the nineteenth century. They lived at a time when few people considered the biological meaning of fossil bones. Nobody knew the true age of Earth, and the majority of the ancient creatures we know about today had not yet been discovered. There were no museums full of dinosaur bones; there were probably never any casual dinner conversations about the physiological parallels between humans and chickens. Darwin lived when the ideas that so captivated me as a kid were just beginning to take shape in academic circles.

For many centuries, people in Europe and elsewhere had taken as gospel (perhaps more accurately, *from* Gospel) that the world had always been pretty much the way it looked right then. But in the late eighteenth century, several thinkers started to question long-held beliefs. The Scottish naturalist James Hutton studied Earth and its natural processes. He is generally regarded as humankind's first geologist. He reconsidered the idea that Earth has always looked the way it does right now. His writing is hard for me to follow sometimes, as it's in florid prose no doubt designed to impress his colleagues, but try this one: "Time, which measures everything in our idea, and is often deficient to our schemes, is to nature endless and as nothing . . ."

I would rewrite his idea in this way: "Time is part of our thinking in everything we do, and often we seem to have too little of it; in nature, though, there is no limit to the amount of time available . . ." This insight led Hutton to realize that the landforms he observed were not the product of a creator clocking in for just six days of work and then heading off for a little R & R. Instead, the geology that Hutton observed and documented was clearly the result of countless years of steady change. He rejected the standard story that there had been a great flood, leaving just a few thousand years of history to create everything we see today. Instead, he deduced that there had been slow, continual change of Earth's surface for eons and eons.

Hutton's idea is called "uniformitarianism," and it was one of the crucial insights on which Darwin built his theory of evolution. Uniformitarianism denotes the idea that the world is uniform, or consistent with one set of natural laws; it connotes another idea that the natural laws we deduce today are the same natural laws that applied eons and millennia ago. It's quite a departure from what Hutton's contemporaries believed (and what creationists today still do). They believed that a creator could change natural laws to suit her, him, or itself. By necessity, then, natural laws and the natural history of Earth could not be uniform. It was to both Ned Nye Boy Scientist and Bill Nye the Science Guy a completely unreasonable point of view. But, we have the benefit of another century of human thought to influence our reasoning.

By the 1830s, the British scholar Charles Lyell elaborated on Hutton's work. I like to say Lyell calibrated the world. He measured how quickly (or slowly) sediments were laid down. He estimated the age of rock layers by measuring them and integrating one age or period of time with another. He was establishing the timescale of Earth. Lyell wrote clearly and with insight into the nature of the enormous spans of time involved. Charles Darwin carried Lyell's *Principles of Geology* with him on his famous voyage around the world aboard the HMS *Beagle.* Several geologists that I know have a copy of Lyell's book on their office bookshelves even today. It stands the test of (deep) time.

Hutton and Lyell were pushing against a strong academic tradition that looked at the world in totally different, static terms. If you visit the Smithsonian Museum of American History today, you'll come across a statue of George Washington by Horatio Greenough, unveiled in 1841. Even as a kid, I thought the statue looked a little odd. I mean, George Washington didn't wear an ancient Greek chlamys (toga thing)—did he? In this statue, he does. So revered were the ancient Greeks that people celebrating the birth of the U.S.'s first president thought it made sense to put an eighteenth-century politician in a minus-fourth-century Greek outfit. As part of that style of thinking, Aristotle's ideas on the relationships of living things persisted well into the age of Hutton, Lyell, Wallace, and Darwin.

Back in that fourth century BC, Aristotle postulated that there is a *scala naturae,* a ladder of nature. In this Latin usage, a ladder is not something you use for climbing. Instead, here, nature's ladder is how things are arranged or displayed from bottom to top. Nothing on the ladder is climbing or descending; each living thing is placed on each step or rung, like books on shelves. Aristotle observed the extraordinary, even exquisite balance of nature and figured that a creator or natural force set each living thing in its exact place, and there they or we each stayed. Everything fits perfectly, like pieces in a vertical puzzle. Along with this arrangement, though, was the idea that things do change with time. Babies grow up to be cowboys (or discus throwers), for example. They have life cycles, during which they grow and change. But in the bigger picture, they remain in their assigned places. These were parts of the perfect ladder of nature.

Taking this perfection into account, Hutton wrote repeatedly that, although he could not help but notice that Earth is ever changing, process, too, was part of a creator's plan. For example: "Nature, everywhere the most amazingly and outstandingly remarkable producer of living bodies, being most carefully arranged according to physical, mechanical, and chemical laws, does not give even the smallest hint of its extraordinary and tireless workings and quite clearly points to its work as being alone worthy of a benign and omnipotent God . . ."

The perspective of Hutton and Lyell began to take hold. By the late 1830s, people were actively speculating on the philosophical and scientific consequences of a very, very old Earth. It roughly goes like this: If Earth's surface has changed slowly over countless millennia, does that mean living things like us have also changed over time? In turn, that might mean that there is no one—to wit, no god—in charge right now. Instead of animals and plants striving for perfection to take their rightful place on nature's ladder, as ancient philosophers had presumed, we are all just blinks in the slowly moving picture of the long, long stretch of time.

Like my grandmother's tenant, many observant people had noted the connections and morphological relationships between different animals and plants. That's how, in the eighteenth century, the botanist Carl von Linné (Carolus Linnaeus, as it is written in Latinized form) came up with his naming scheme for connecting one type of organism to another. Any living thing could be categorized in a hierarchy each level of which was a binary choice. To place something in a catalog of living things, the naturalist just had to decide, is it animal or plant? Does it have mirror-image symmetrical leaves or long flat leaves? Is it woody or smooth? It's like a game of Twenty Questions. The Linnaean system further encouraged naturalists to think deeply about the relationships between living things. Linnaeus was so influential that the Linnaean Society is going strong still today.

By the time Darwin was born in 1809, a number of naturalists were starting to explore how different varieties of living things were related and how they could change over time. Earth seemed old enough to allow such changes to happen, but nobody knew how one species could become another species, no matter how much time was involved. One person came close but got the main idea wrong—you'll read a whole lot more about him in the next chapter. Then along came Darwin and Wallace and . . . here we are. Later, people like Estonian explorer Karl Ernst von Baer and philosopher Johann Goethe explored the ideas further; later still came Niles Eldredge, Stephen Jay Gould, and many others, each adding another element to the scientific story.

Darwin's idea of evolution through natural selection sparked a broader fascination with the idea of competition in the natural world. I've got a good deal more about this in chapter 8. The idea of competition inspired "social Darwinism," which looked at competition among human populations (often in racist ways that were unrelated to what Darwin actually wrote). Meanwhile, Darwin himself got to thinking about populations: not just human populations, but populations of every species observable in nature. He saw that there is a spectrum of variation between species that comes into being slowly, tiny change by tiny change. He realized that variations happen naturally when an organism reproduces. He realized that populations of species compete for resources. And he realized that the traits that are inherited, which benefit the organism, have a greater chance of showing up in that organism's offspring, providing the engine that drives evolutionary change.

Although Darwin and Wallace came upon the idea of evolution at almost the same time, I can see why we associate the theory of evolution solely with Darwin. (I'll tip my hat to Wallace in a later chapter.) Darwin's book is amazing. *On the Origin of Species by Means of Natural Selection*—that's the full title—includes dozens of diligent observations and experiments that Darwin personally conducted. It is also beautifully written. He leaves the reader to draw his or her own conclusions as to whether or not the theory of evolution is the real deal. Just for example: "On the view that each species has been independently created, with all its parts as we now see them, I can see no explanation. But on the view that groups of species have descended from other species, and have been modified through natural selection, I think we can obtain some light . . ."

In his writings, Darwin makes it clear that he cannot state whether or not there is a creator in charge. The idea was impossible to prove or disprove then, and it still is today. But what emerged from Darwin's steady investigations was a new view of the world that can be appreciated and understood on its own terms. Perhaps there is intelligence in charge of the universe, but Darwin's theory shows no sign of it, and has no need of it. The exquisite variety and balance that we see in nature is a result of nature itself.

I know that this realization still bothers a great many people. For me, it is stunning and uplifting. After 2,400 years of speculation, humankind has finally uncovered this fundamental aspect of nature and our place among living things. Just think what other equally revolutionary discoveries lie right around the corner.

7

LAMARCK AND HIS NOT-ACQUIRED TRAITS

By the time Darwin came along there were and had been a great many naturalists—what today we might call biologists—who had been paying attention to the similar shapes and functions they observed among plants and animals. Philosophers had been wrestling with the origin of life since the heyday of the ancient Greeks. Almost all had speculated on how life began, and how living things became so obviously interdependent. They observed and documented patterns in nature. Big fish need little fish. Squirrels need trees. Humans need to eat, and nature provides the groceries. But how did the great variety of living things come to be; how did we all get here?

Living things were generally believed to have a soul or a metaphysical quality that was passed from parent to child and even from tree to acorn. In the minds of nineteenth-century naturalists, permanence had been there from the beginning; change was not built in. Yet they suspected that a mechanism for change did exist. How could nature or a deity cause so many forms to arise or be created? Did a creator imbue every seed with a soul? Expressing it in the words of the nineteenth century: How could what they called "homogeneity" become "heterogeneity"? Darwin's descent by natural selection was a result of not only a great step in thought, but also the sidestepping of a great many ideas that missed the mark.

A number of Darwin's predecessors latched onto some aspects of natural selection without getting the fundamentals right. And nobody did a more spectacular job mixing correct and incorrect than Jean-Baptiste de Monet, who went by his courtly, inherited French noble name of Chevalier de Lamarck and is commonly called simply "Lamarck."

In mid-eighteenth-century France, Lamarck speculated that when animals and plants exercise certain organs, those that use other organs regularly and vigorously would not only enhance the function of said organs, but would pass on the tendency to have an improved or strengthened organ to their offspring. Apparently, he observed that blacksmiths had muscular arms and shoulders, as one might expect from a lifetime of hammering on anvils. He expected that a blacksmith's kids would inherit those arms, passing along big shoulders, or perhaps the ability to develop big shoulders. From there, he drew a bigger conclusion about the bigger system in nature.

It's not hard to see why Lamarck thought that way. We are all familiar with people who follow in their mother's or father's footsteps. If you are the son of a blacksmith, you would grow up knowing more about metalworking than the average person. When you come of age and seek a trade, it's reasonable that you'd have a leg (or a hammer) up on other would-be blacksmiths. It must have reinforced the conclusion in Lamarck's mind. Today, there are a great many major league baseball players who are the sons of major league players. It probably has something to do with being exposed to baseball culture and the details of the rules in addition to inheriting the right physique for the game; but you can see how these factors working together might result in a son following in his father's footsteps. You might think that the more you did something—the more you swung a bat or a hammer—the better you would get at it. Along with that you might conclude that the children of a horseshoer, a barrel maker, a bat swinger, or a ball thrower would tend to inherit these abilities. If that is true of people, why not animals as well?

Seeking to link cause and effect, Lamarck speculated that this supposed ability to change or modify the traits passed on to offspring was the result of a *complexifying force*. If an animal wanted to eat certain leaves, for example, it might develop the right kind of teeth for it, and then pass that beneficial new trait to its offspring. This came to be called the inheritance of acquired traits. It would be a tendency or agent in nature that helps the successive generations build on beneficial modifications that resulted from a previous generation's effort.

Lamarck's speculation was a helpful means to gain insight into the ways in which species change. He was grasping to understand the means by which an organism could gain complexity or specificity and efficacy as he, she, or it reproduces. For me, the iconic example is the giraffe (*Giraffa camelopardalis*). Imagine being a central European thinker and coming upon a giraffe for the first time; because of their unassuming ways and their seeming amicable outlook on life, giraffes are just fascinating. Inevitably you would wonder: How did they get those long necks? Why don't they have regular necks like dogs, cats, and cows? After all, we do not observe giraffes actively stretching their own necks as they grow up. They are just born that way. If you cut the tail of a mouse, its offspring still have tails. If a big-shouldered family stops working as blacksmiths, they still have big shoulders. Lamarck's ideas did not stand up to scientific scrutiny.

We know now that a giraffe's neck, like all its physical attributes, is controlled by genes, and living things in nature cannot alter their genes. All organisms—sea anemones, fireflies, giant squid, miniature poodles, and humans—have to play the genetic hand they're dealt. What Darwin realized, and what Lamarck missed, is that the complexity emerges slowly through a whole population, not quickly within a single person or animal. With that said, researchers have recently discovered an intriguing twist to the story. Under the right conditions, and up to a point, inheritance *can* work the way Lamarck believed it did. Although genes themselves do not change on their own, the way that those genes are activated can change within a single individual's lifetime. This phenomenon is called epigenetic change—changes coming from the outside.

In the not-so-distant future, there may be another way that your genes can change. Scientists are working on gene therapy—ways to alter your DNA to eliminate or obviate a disease, to make adjustments, or to incorporate what we hope are improvements to our genes. Someday it may be possible to make modifications in the so-called germ line, which would result in that new DNA being passed on to your kids. Creepy or fabulous? Crazy or ethical? The possibility of genetic modification reminds me of the need for a scientifically literate electorate. Please stay tuned and vote!

Keeping all of this in mind, let us once again consider the giraffe in its natural habitat. Having traveled in Africa a couple of times and observed giraffes in nature, I can tell you that you don't have to be the world's foremost authority to see that giraffes eat leaves on branches pretty high off the ground. They use their necks to get to vegetable matter that other animals would have to work quite a bit harder to get to. If you were a cat, you could climb out on those branches and gnaw away. But it would be a lot more work. Oh, and cats generally eat other animals rather than delicious acacia tree leaves. Giraffes have another fascinating feature that I didn't notice, until it was pointed out to me. They have very tough tongues and lips. They can just grab onto the thick part of an acacia tree limb and slide their mouths right down the limb to the thin end, stripping all the leaves off as they go. Here's the thing: African acacia trees are loaded with thorns. You and I cannot grab an acacia branch with our bare hands, let alone our bare tongues. Yee-ouch! But giraffes can.

Thinking as Lamarck had thought, you might conclude or presume that the giraffes that we see today got their long necks by just reaching. You might think that just by stretching their necks to get food, giraffe necks would naturally get longer—and so would the necks of their offspring. But no. Darwin reached the correct answer: The ancestors of our modern giraffes, who happened to have slightly longer necks than their contemporaries, were able to reach just a little bit higher on acacia trees than the other members of their giraffe herd, or no kidding, other members of their tower. (It's like school, as in a school of fish. Only this is a tower of giraffes.) The giraffes with longer necks were better—just a little bit better—at getting enough nutrition. So, they were just a little bit better at having babies, too.

The evolutionary pressure to have longer necks was probably stronger when food supplies were scarce. Imagine a drought on the savannah, the African landscape that a European or American might describe as somewhere between a forest and prairie. During the drought the trees that survive probably have fewer, smaller leaves that don't carry as much water as they do when there's been plenty of rain. In this situation, all of the animals that eat acacia tree leaves can reach the leaves on the lower branches. But only the giraffes who happen to have the slightly longer necks can keep going and reach and eat the higher leaves that other animals and other shorter members of their tower can't reach. So when the scarce leaf supply is gone from the lower branches, the tall members of a tower get more food and are more likely to have healthy babies.

Now, imagine a drought like that happening almost every year for, say, ten years. Like North America, Africa is subject to climate patterns associated with El Niño events in the western Pacific Ocean. Such patterns can last for years. Over the course of a few seasons, a tower (a population) of giraffes might find little to eat. In that case, only the tall would survive. The pressure would be high. Instead of giraffes that were a little bit taller doing a little bit better, the taller ones would be the only ones to make it through the drought. The shorter members of the group would, in just a few years, die out. Their genes would be eliminated quickly.

This idea illustrates, in simplified fashion, how changes in the environment can select for well-suited genes surprisingly quickly. You can't stretch your own neck to give your kids long necks. You have to have long-neck genes (or longer-neck genes), and they have to make it into your offspring. Poor Lamarck, smart as he was, did not see how the process really worked. But as modern observers, we have to give credit to Lamarck for even taking on the problem, for even thinking about it.

While we're talking about giraffes, there is another remarkable and vital point to be made about evolution and the survival of the "good-enough." It is an unfortunate linguistic happenstance that "survival of the fittest" sounds so good, because random natural variation does not produce perfectly fit individuals, nor does it need to. Evolution is driven by the idea of "fits in the best," or "fits in well enough."

When we look at the anatomy of a giraffe, we come across a great many surprising and interesting features. First of all, although a giraffe has what seems to us a pretty long neck, a giraffe has seven vertebrae, just as you and I do. Her or his neck is, in a fundamental sense, the same as ours. This is evidence of a common ancestry. Somewhere back in time, there were vertebrate mammals (those with backbones) that gave rise to both giraffes and to us. Seven vertebrae are not very many for the giraffe's long neck. Having such few, large bones limits the animal's flexibility. But evolution constrains us all to work with what we've got.

Along this line, the nerve that extends from your brain to your voice box, your larynx, runs down from your brain and past the larynx. It goes right by your larynx like the pavement of a big city beltway. This same nerve runs around an artery near your heart, and then back up to your neck, where it connects to your larynx. It really does. The same is true for a fish, where the nerve from the brain to the gills takes a pretty short route. But with generation after generation, certain animal necks got longer. Gills changed so that they could take in oxygen from the atmosphere rather than take in oxygen dissolved in water. This same nerve kept running the same route. Down from the brain, around a heart artery, then back up to the larynx. That's another consequence of evolution: Every generation can only be a direct modification of what came before.

In a giraffe, it's wild. The nerve runs from the brain down to the animal's heart, which is in its chest, just like yours, then back up to its voice box, the larynx. If you were to sit down and design a connection from the brain to the larynx, you'd make it just 5 centimeters or so (2 inches). But because animals like us and giraffes came from ancestors with the same kind of nerve wiring, we end up and giraffes end up with this arrangement, which seems quite odd at first. After you mull it over, it's just what you'd expect from fish, from yourself, and from giraffes.

The same must be true of a giraffe's tongue and lips. Each generation of proto-giraffe, the ancestors of modern giraffes, had successively tougher tongues. A tough tongue enabled you, as a proto-giraffe, to get a little more to eat from the leaves of high branches. Eventually, a generation of giraffes was born that could eat high and thorny acacia leaves.

To drive this idea home, join me in a little thought exercise. Imagine a bicycle. Now, imagine a two-wheeled hand truck or a two-wheeled cart, the kind you see city dwellers bring to the grocery store. Now, imagine making changes to the cart or hand truck so that it becomes a bicycle—but making those changes step-by-step, in accordance with evolutionary principles. You would have to stretch the cart into a parallelogram shape or something similar, so that the wheels were one before the other rather than side-by-side. The wheels would have to be bigger. The tires would have to hollow out and become filled with air. You might have to modify the axle to become a chain. Maybe the handle that goes over the top would somehow have to become the top tube of a bike frame. And every bit of the cart would probably have to be thickened to become a bike frame able to support the weight of a human bouncing over a rough road.

The crucial evolutionary requirement is this: At every stage, with every change you make, the whole thing has to still be functional. It has to still roll. It has to still be drivable or take-to-the-store-able. Otherwise, the cart or hand truck would die out. If at any point it didn't roll, or couldn't be steered in some fashion, or if there were no practical way to keep it in balance, you'd have to abandon it. You'd just leave it by the side of the road to rust to dust, and try again with one of your other related designs. Inevitably, you'd have to keep a lot of the original design and make changes in an incremental way. That's how things are in nature, where there is no deliberate designer who can take things apart, redesign them, and put them back together if they don't work. Instead, every step has to be "good enough." Every generation has to survive lest that species or type of organism will disappear from our world.

That is why giraffe necks are so similar to our necks and to dog necks and to horse necks. If we look closely, our necks are similar to fish necks. We are descendants of a common ancestor way back in

Earth's history. Configuring necks this way is almost certainly not how a designer or engineer would build the world. But the details all make perfect sense once you embrace the idea that evolution does not work the way a human designer or engineer would.

Evolution happens as each generation of living things interacts with its environment and reproduces. Lamarck got at least that part of it right. Those natural designs that survive to reproduce pass on their genes. Those that don't successfully reproduce disappear; their genes disappear as well. It's survival of the hang-in-there's, or the made-the-cuts, or the just good-enoughs.

8

MY PROM AND SEXUAL SELECTION

Being a nerd, I did not anticipate going to my high school prom. Nevertheless, I did. I was driven to do so, apparently, by the shape of Leith's legs, a (clearly) female classmate. This fascination with sex is, near as anyone can tell, not something we get to choose. Our ancestors bequeathed it to us. It's another one of those deeply shared evolutionary traits. It's a drive we cannot disengage.

Along this line, I cannot help but recall a day at a beach in Delaware, when my mother's first cousin Monique showed up to sun herself. I was about seven at the time. My mother's mother was French, so my first cousin once-removed was French, too. She had a certain European flair. And there was another thing: Monique was in her twenties, and wearing a bikini. (I am not skilled enough to have prepared a sketch.) I remember the grown-ups staring at me, because I was staring at her. The thing is, I distinctly remember that I did not know why I was staring at her. Of course looking back, she was, as it is expressed in modern parlance, a total babe. But I did not have the linguistic skills to express that, nor did I feel anything like what I would later feel as a teenager. I was just stuck staring. I take this as incontrovertible evidence that our brains are set up to support or carry out sexual selection, without our even knowing it.

Sexual selection is the second fundamental idea in Darwin's theory of evolution, ranking next behind natural selection. Sexual selection is the process by which organisms of the same species select genes to be passed to subsequent generations. It is what drives so much of what so many of the species on Earth do all day, and all night.

Natural selection in the general sense is the interplay between organisms and their environment. Nowadays, a century after Darwin, we might describe the process as the interplay between organisms and their ecosystem. Slightly better suited organisms outcompete the not-quite-so-well-suited organisms. This happens when random processes produce genes that happen to fit in well with the environment and ecosystem extant at the time. It was Darwin's greatest insight, and still forms the cornerstone for the modern understanding of what drives evolutionary change.

But along with the interplay between individual and ecosystem is another interplay between individual and other individuals of the species. They are competing with each other for energy and opportunities that enable them to reproduce. For plants, the fundamental resources are sunlight and nutrients in soil. For little fish, it might be zooplankton, tiny animals in the sea. For big fish, it's little fish. For you and me, it's food and water. But from an evolutionary standpoint, no amount of sunlight, fertilizer, nutritious food, or cozy blankets is enough. Organisms have to pass on their genes in order to have successors—in order to keep their genes in the gene pool. To that end, plants and animals go wild.

There are celebrated passages in *The Bible* about the lilies of the field. The authors of the Book of Matthew remarked that these beautiful flowers neither toil, nor do they spin yarn to be made into clothes. It's a passage encouraging the disciples of Jesus not to worry about Earthly things, specifically what exactly they'll be wearing while they're out proselytizing.

Lovely as that passage may be, *The Bible* missed an important point here about nature and evolution, specifically about sexual selection. In fact lilies, like every other sexual organism on the planet, work pretty hard to produce a means to mate. If you haven't already, stop and think about how much energy a plant puts into creating a flower. In general, a green plant such as a lily, rose, hickory tree, Ponderosa pine, or bull kelp, has leaves, needles, or fronds to collect sunlight. And in general, the other structures such as stems, trunks, or stipe serve to support the leaves in an efficient or efficient-enough fashion. What else does a plant do besides look to soak up light? The answer is simply: make more plants, which is not easy.

Plants go to great lengths to reproduce. It takes a lot of a lily's energy to produce flowers. It takes a great deal out of an oak tree to make thousands of acorns. In that case, the tree is, in turn, counting on squirrels to forget where they hid a few acorns, so that a new oak tree might grow nearby. Apple trees and orange trees go to all kinds of trouble to grow fruit, so that some guy like me or my local Los Angeles "citrus mice" (rats) will wander off with a piece and spit out the seeds on suitable moist soil. Palm trees grow coconuts the size and toughness of cannonballs, so that they can float their seed to another island. Just consider how much less energy it would take a lily or a cornstalk to grow and thrive if it didn't have all these seeds to sow.

And there is more. It's not just that these organisms are growing viable seeds. It's also that these organisms are growing structures such as flowers, pistils, stamen, eggs, and pollen to get a mix of genes, before the seeds are sent off. It's all sexual.

A rosebush has woody canes to give it structure. It has thorns to discourage animals from climbing on the canes or perhaps using them for a nest. Producing canes takes energy. But just look at the resources and energy rose plants expend creating elaborate blooms and hips (rose seeds). They produce attractive flowers not to avoid germs or to make it through a cold tough winter, but to get their genes mixed with the genes of other individuals that are selected on account of sex. They do it to attract pollinators, like bees and birds, that stop by for some nectar and carry some pollen when they fly away.

Oranges are so wonderfully sweet and appealing, and sheep wool is so soft and toasty, that it's easy to imagine how our ancestors could believe the whole setup was put there just for us. But this is clearly not the case. Ecosystems come to be over enormous amounts of time. We show up for just a few decades or generations. Our ancestors, leastways the ones who labored over those passages in *The Bible*, missed the idea that the reason it all seems to fit together so well is that it's been developed, from the simple to the complex, over a span of eons. And sex made it happen quite a bit faster, or a bit more efficiently, than it would or could have otherwise.

Once it arose, at least 1.2 billion years ago, sex became popular among living things—I mean, it's everywhere. We can speculate about how sex began. Under a microscope, we can readily observe bacteria exchanging genes across what resembles a microscopic string or thin tube. It's called a "pilus," Latin for "hair." It's not hard to also imagine bacteria exchanging just portions or tiny bits of genes. Next, imagine one bacterium sending to another small bits of genes (bits of bits) that are lighter in weight, and sending the material very quickly, while another bacterium favored sending larger heavier molecules more slowly. As we observe them now, bacteria only share one to one, and there is always a donor and recipient. Suppose though, that one bacterium sends a few genetic molecules through its pilus, and then a few minutes later the recipient bacterium turns it around and sends some other genetic information to that first donor through its own (different) pilus.

In primordial times, a situation apparently developed in which one bacterium sent many smaller bits in exchange for another bacterium's heavier larger bit. This practice apparently gave the microbes that settled into the two roles an advantage. Each must have performed its job a little better than other microbes that continued the practice of having both organisms on either side of the exchange sending longer or intermediate-length chains of genetic code. Specialization apparently introduced efficiency. One thing led to another and another, and these primitive Earthlings invented sex. Just looking around us, we can see that this exchange of many small pieces of genetic code for one larger piece works well. (Why it works well is less immediately obvious; more on that later.) Sex must give some living things an edge. Otherwise it, and we, wouldn't be here.

Once established, sexual selection is easy to understand. It enables a species to select within its own kind, and individuals within that species to compete in the open world, the open ecosystem, if you will. Another way of looking at it is that sexual selection is a second filter upstream of natural selection: Before offspring are introduced into the world to see if they are good enough to produce their own offspring to advance into the genetic future, the parents have to select each other. If they don't, there are no offspring to be introduced to the world.

The speed of sexual selection also contributes not only to the remarkable diversity we see in nature, but to the complexity of organisms as well. When offspring carry new combinations of genes that can take advantage of resources in the environment faster than other organisms in their ecosystem, they are bound to succeed. If they take in resources such as nutrients and water more quickly or more efficiently than others around them, it is reasonable that their genetic innovations will incrementally become more complex, because they can support more complex genes.

If you're like me, and I know *I* am, you might wonder why we have only two sexes. If two sexes are better than one from an evolutionary diversity-generating, germ-fighting, complexity-adding, outcompeting standpoint, why not have three sexes, or four sexes? Then, you—and all living things—might be able to innovate genetically like crazy. As a first hypothesis, this makes sense—at least it makes sense to me. But, we must keep in mind that evolution by means of both natural selection overall and sexual selection specifically can enable an organism to become more complex only from where an organism is at any moment. Getting multiple microbes of multiple generations to interact simultaneously may have been too rare or provided too little of an advantage over a faster one-to-one scheme, so we have just our two sexes. The next generation of an organism can introduce innovations only from the complexity or level of innovation its ancestors already had. It's not the survival of the absolute fittest multiparent scheme imaginable. It's survival of the good-enough.

Although most of us Earthlings presume to deal with two sexes, in the world of fungi, things are somewhat different. They interact two at a time, all right. But they have what are currently called mating types. They can be sexually compatible with many other individuals, many other mating types of their species that are in a sexual sense quite different. In this sense, an organism like the split gill fungus has 28,000 different sexes, i.e. 28,000 mating types. For us two-sex creatures, it's fascinating. Putting the unusual gene-sharing practices of the domain of fungi aside, the other sexual creatures like us work the problem with just the two—male and female.

No doubt there have been times in your life when you wanted to make multiple copies of something you created. These could be fence pickets, birthday party invitations, or laser gyroscope navigation systems for small jets. As you set about the task, you are probably going to make the second copy from the first. One is going to become two. The same holds true in nature. Let's say molecules used chemical energy from their surroundings to make copies of themselves. When they made a copy, it's just that: one copy. Then that copy could make a copy, and so on. We're talking about the molecular level here. So, the design that emerged over billions of years—not millions, but billions—is a molecule that splits itself in half.

In the case of the primordial gene-sharing microbes, the ones that apparently led to the invention of sex, they were only able to share with one other microbe at any one time. We do not observe multiple gene-exchange partners in microbes. And we observe only two sexes today.

Nature ended up with the deoxyribonucleic acid molecule, DNA. So binary splitting is favored or inherent in the process. With that, binary combination is as well. This is what happens in natural systems without us involved. It's also what happens in human systems. When we play the World Cup soccer matches, we play two teams at a time. Any athletic or bridge tournament involves groups of two teams. It's hard to imagine a system that works any other way. If you have three teams on a field, it might work for a while. But pretty soon, somebody teams up with someone else and the competition becomes one-on-one, or two total at any time.

I'll grant you that we can have more than one individual or more than one team in a horse race or a poker game. But if this whole sex business started with microbes exchanging genes one-to-one, it's a hard system to undo. In other words, once molecules started reproducing by dividing, it would be difficult indeed for a third team or molecule to develop and compete with a binary system. Perhaps there is an ecosystem on a planet orbiting another star—heck, perhaps there's one on Jupiter's moon Europa—with a tripartite sexual system. It's a concept that Isaac Asimov explored elegantly in his novel *The Gods Themselves*. But a binary setup seems to be what happens when we let nature sort it (or her) self out.

Speaking of two teams, books have been written, operas composed, game shows created, and countless news stories produced that deal with the battle of the sexes. That men and women have such difficulty getting along sometimes, or perhaps most of the time, would be a mystery, if it weren't for evolution.

We are all driven to select a mate. Those of us who don't do not pass our genes on to the future. A human female is apparently driven to select a mate that will provide for her and her offspring. If nothing else, such behavior or motivation would be consistent with the tenets of sexual selection. A human male selects a mate that is, by his reckoning, well suited to carrying his genes forward. The female has to make her genes appear valuable by "playing hard to get," as the old saying was and is so often said. But really . . . all things in moderation.

Look around. So much of what goes on in our society is motivated by the process of sexual selection, and there are many subtle and not-so-subtle things that affect that process: There's mascara. Expensive watches. Amazing shoes. Sports cars, perfume, skirts, ties, jeans, boots, and on and on. Now, compare us to everybody else. By everybody else I mean dogs and cats and lions and tigers and bears . . . and squids and whales. All the other animals around, and all the plants, have seasons to their mating. We humans don't seem to. When it comes to our babies, birthdays are pretty well distributed around the calendar. Why is this? Why our species-wide sexual overdrive?

One of the leading theories is that it's an artifact. It's a result of being good enough in the process of evolution. The battle of the sexes is so strong perhaps because all of us ended up with big brains (compared to other animals—and my old boss). These brains make us aware of our place in the scheme of things and that somehow leads to doubt and altruism and loyalty and the ability to so easily do the wrong thing. To keep us from not getting around to mating, our sexual selective processes are turned on 24-7. Wherever it comes from, it works. This is to say, although we are continually stymied by broken hearts, loyalty, disloyalty, professional concerns, familial criticism, and other distractions, the human species reproduces at a prodigious rate. In ten thousand years the number of people on Earth has exploded, from a few million to more than seven billion. Apparently, sexual selection left us with our need to engage turned up to eleven.

And so I, a victim of evolution, did manage to go to the prom. It was opportunistic; I admit it. My date spent a lot of time with the older guys, guys from a class or two ahead of ours. When it was prom time, they were all in college and not available. I asked her out, and she agreed to come with. Apparently, neither one of us could help ourselves. I saw her at our reunion recently and she is still, as it is expressed in modern parlance, all that. I guess it's seared in the proteins that create memories in my sex-driven brain.

9

THE TALE OF THE RED QUEEN

In the previous chapter I discussed when sex evolved, and explored how sex evolved, but you may notice that we (or I) didn't really address the "why." Why does anybody—I mean why does any organism—have sex? The amount of energy we put into it is crazy. We spend billions on lipstick and hair product. About every third advertisement in any medium is for a car—one with plenty of sex appeal. We get our pants pressed, our nails done, and we work hard to smell nice—all to attract a mate, one that will enable us to have sex and offspring. Why do we go to all the trouble of attracting a mate? Why not just get this literally vital task done on our own? Why not just let our lips be lips and drive around in gray, featureless blob cars?

It's not as if sex is the only way to reproduce. We humans could, for example, just split ourselves in half, DNA, bones, muscles, brains and all, like any self-unaware, or perhaps self-respecting, bacterium. The separate pieces form new membranes and boundaries as they separate. A replica of the DNA from the parent is built on each side—I mean, inside each of the two individuals that emerge from the original. If it's hard to imagine a person splitting in half that way, try picturing a genetically identical baby budding directly off of a mother or a father. There's no reason why it couldn't work in the big picture, yet here we are instead surrounded by all of those made-up lips, polished nails, fancy fragrances, gym memberships, sporty cars, and the like.

And those are just the obvious examples from our human experience. Much more important, from a planetary perspective, are the billions of other species here on Earth that use energy from the Sun and the soil to build flower petals, pistils, and stamens—to build bioluminescent appendages—to grow fruit on their limbs just so some jerk like me will yank it free, cart it off, and spit the seeds out someplace, where they might find friendly soil and a place to grow some offspring fruit of their own. Why bother?

Salmon swim and swim. They spend most of their lives eating other fish with the goal of finding a fish-mate that they can swim upstream with, lay a pebble bed, fertilize with milky sperm, and die. Why bother? You can try not to make jokes about it: A pregnant elephant takes up a lot of room, but getting an elephant pregnant looks like a good bit of difficulty. Why bother?

Let me start by saying as of this writing, no one is absolutely sure why organisms like you, me, armadillos, and trees have sex. Broadly speaking, sex produces offspring with a new mix of genes, a mix inherently different from the parents. It may provide more chances for innovations, which might lead to a more successful successive generation. The new mix of genes might weed out genetic errors.

But we do have an outstanding theory about why sex is useful, a theory that places sexual reproduction within the larger picture of the theory of evolution. It is survival of the fittest, survival of those that fit in the world, specifically the ecosystem that is best. It's competition. And one's chief competitors are seldom other troublesome large animals: Humans really have very little trouble keeping up with and living around lions and tigers and bears. Instead, our most troublesome bad guys are germs and parasites. These are what can kill us or disable us to the point where we cannot produce or care for offspring.

We're not the only organisms with this germ and parasite issue. If you lick your lips, you might ingest right around 1 million viruses that specifically attack bacteria. By long tradition, they're called bacteriophages, or just phages for short. (*Phage* is from the Greek word for eating. Phages eat the insides out of bacteria.) In other words, even bacteria, these relatively uncomplicated single-celled organisms, have great difficulties with phage viruses, whose sole function in life, or along the border of life, is to hijack a bacterium's metabolism and use it to make copies of themselves.

A striking feature of phages is how specific they are. That is to say, a specific phage only attacks a specific type of bacterium. The surface of the phage identifies and sticks to a specific protein on the specific bacterium. The main defense a bacterium has against a phage attack is to somehow modify or reconfigure the protein pattern on its outer membrane. Now, individuals cannot change themselves, as such. Instead, their descendants, their offspring, can have modifications as their DNA is replicated. Random changes may or may not help them resist a phage. Keep in mind that we're talking about bacteria. They may not change very quickly, from one generation to the next. But they reproduce like crazy, doubling over just the course of a few hours for most species, so bacteria produce almost uncountable numbers of descendants. In those lots are bound to be new configurations of imperfectly copied genes, some of which will have the ability to resist or just not be identified by phages that could have killed their ancestors.

It is this fast replication of bacteria, viruses, and other not-too-complicated parasites that can cause us big organisms trouble. The germs are out there reproducing and chancing upon ideal protein-pattern-attachment genes, which are well suited to infecting us. Meanwhile, big organisms like you, me, and redwood trees cannot reproduce over the course of a few hours. It takes redwood trees centuries. It takes us months and months, then years more to get ready to do it again. So to stay in the evolutionary game, we, and especially our ancestors, have (had) to come up with a different scheme, or we would never have been able to keep up with all the phagelike viruses out there all set to take our cell metabolism and turn it against us. Put more accurately, our distant ancestors did indeed come up with a different scheme; otherwise you and I wouldn't be here to ponder the question(s).

The key to fighting germs and parasites seems to be sex. At one level, this may bring you down. All the lipstick, high heels, hair products, salary seeking, sports cars, and weightlifting seem to be a result of germs. But then, so are art, and music, and good cooking. By having sex, organisms like dandelions, sea jellies, perch, parakeets, and termites can stay ahead in the game of life just enough to have offspring that succeed in producing more offspring in a subsequent season.

By relatively recent tradition, this is called the Theory of the Red Queen. The charming sobriquet comes from the fictional Alice of Lewis Carroll's books *Alice's Adventures in Wonderland* and *Through the Looking-Glass.* In this story, Alice has an encounter with the Red Queen. (By many accounts, Mr. Carroll smoked some form of marijuana from time to time, and perhaps enjoyed a glass of wine or two.) Somehow the Red Queen is a sort of hybrid chess piece/person, who slides along, on something akin to a Chess Board of Life. So when one is with the Red Queen, her whole world is moving . . . somehow. So, Alice is constrained to run like crazy to have a conversation. This is the Red Queen's day at the office or day at court. Alice remarks, "Where I come from, if you run all day, you end up somewhere else."

The Red Queen, as she raises her royal-chess-piece-person eyebrows, remarks, "Why, that seems like a very slow sort of country. Here it takes all the running you can do to keep in the same place."

Apparently, the process of evolution that you and I, and every other living thing on Earth, are caught up in is like the land or country of the Red Queen. In evolution, we have to run constantly; we have to continually come up with new combinations of genes to keep our genes in the game. To be successful as a living thing, you have to have offspring, who have offspring, who have offspring. Rest assured, your family did, or you wouldn't be here. As troubling as it may seem, your parents had sex—at least once. If you have brothers and sisters, more than once . . . One shudders to think of it.

Like any scientific theory, we can use the Theory of the Red Queen to make predictions of phenomena we see in nature. When I participated in a debate with the creationist Ken Ham, I used the example of the small fish called topminnows. These fish are remarkable. When times are tough and the pickings of opposite sex fish are slim, these particular topminnows (*Poeciliopsis monacha*) can produce eggs that develop into fish, without the egg having been fertilized with another fish's sperm. This is known as asexual reproduction. Of course, boy topminnows and girl topminnows also get together to reproduce sexually by having regular fish sex.

The topminnows in question live in Mexico, in rivers. When it rains and rains, there are plenty of pools in which the fish can make their homes. When the weather dries up, the pools get separated by tracts of dry land. These fish are attacked by a parasitic flatworm, which in English we call the black spot flatworm. In the isolated populations of fish, the ones that were forced to reproduce just by themselves were subject to more black spots than the same species of fish that had enough prospective mates around to enable them to reproduce sexually. Wait, there's more. There was also a gradient to this disparity:

The populations that had a higher percentage of asexual reproducing individuals also had a higher percentage of black spot flatworms.

There's more, more: The researchers, Bob Vrijenhoek at the Monterey Bay Aquarium Research Institute and his colleagues, found a population in a certain isolated pool, where the sexually reproducing fish had more parasites than the asexual fellows or females. They discovered that the sexual fish had inbred; they had had sex with their brothers and sisters more than outsiders. So the random mutations introduced by the asexual reproducers were outpacing the sexual ones, which was surprising. Since there is no shortage of these fish and no shortage of such pools, Vrijenhoek introduced some outsider sexually reproducing topminnows into that pool's particular mix. A season later things were back, as one who acknowledges the influence of the Red Queen would expect. The asexuals were more infected than the sexuals.

It's a wild world out here where we live. But these fish are a remarkable example of a scientific theory making predictions that come true. In this species, we could compare directly the effect of a single type of parasite on a single type of fish. They're wonderful special fish, the topminnows, because they reproduce in two different ways, and we can observe them from one season to the next. We do not have to wait decades or centuries, as we would if trying to study humans or giant sequoias doing this.

This is one of the reasons I get such joy from studying evolution. This kind of science is amazing and sexy.

10

DOGS ARE ALL DOGS

People love dogs. This is, I hope, the least surprising sentence you will read in this book. I myself have had long discussions with my dog friends, and by that I mean my friends who are dogs. I will admit that these discourses were largely one-sided. When dog people get together, they often ask each other what or which kind of dog does each person has. They're talking about breeds of dogs: Collie, corgi, Labrador Retriever, pit bull, or poodle. Generally, I love 'em all, because dogs are all dogs. They are a great walking, barking definition of what a species is. Evolutionary relationships are defined not by looks but by what's inside. If a male and a female can hook up and produce offspring, then by the simplest and most meaningful definition they belong to the same species.

Put another way, I love all of you dog lovers, but I have to spoil your fun a little with a fundamental truth. There is, in an important evolutionary sense, no such thing as a specific breed of dog. If a Great Dane has sex with a dachshund, you get a dog. If a Standard Poodle has sex with a Jack Russell terrier, you get a dog. If a mutt has sex with a so-called purebred, you get a dog. You don't get anything else. All dogs are descendants of common ancestors. So, when we all enjoy the kennel club's show, which is divided or organized by breed, we are participating in a ritual that is, at an important level, arbitrary. There are gradations, or there is a spectrum of dog types or breeds. The word *purebred* is something we can define by counting generations back in dog-sex land. But it is not an indication of species or anything special, really.

Our modern domestic dogs are direct, and I mean direct, descendants of either wolves or a common ancestor a step earlier than wolves. Since we have access to machines that use carefully developed reagents to unwind and meticulously count the sequence of nucleic acids—the coding chemicals that are assembled like rungs on a ladder—in an organism's DNA, we can compare directly the DNA of a wolf to that of a New Zealand Blue dog, for example.

Even more remarkable is a set of experiments led by Dmitri Belyaev on a Soviet fox farm during the 1950s. He and his colleagues gained access to a pack of silver foxes, which are prized for their fur and are closely related to wolves. The researchers observed the foxes and offered them human food. The individual foxes that were less inclined to run away as humans approached were rewarded with food. These foxes were selected and allowed to breed, male to female. In just a few generations, the researchers had bred fox pups that were not skittish around the staff. These newly bred fox-dogs or dog-foxes now had an affinity for people. They wagged their tails in joy, whimpered for attention, and licked the experimenters to express affection; some even developed floppy ears, which are scratchable.

The same process must have happened during the domestication of wolves. They were willing to protect our ancestors as though they were one of their own. And this was an apparent result of our providing food and shelter, nice dog-wolf housing. All modern dogs are very nearly wolves, genetically speaking. We humans have decided which wolf-dogs would be allowed to have sex, and so selected the wolf-dog traits that we wanted: friendliness, snuggliness, playfulness, and protectiveness.

Darwin was quite diligent about the business of distinguishing which living things are logically to be grouped with others—specifically which living things were of the same species, able to breed with one another. He observed that what a great many of his contemporaries regarded as different species of roses were, in fact, just slightly different varieties of roses that were easily bred together, or crossed. Darwin's distinction confused a great many people. The roses might look quite different in shape and especially color from one generation to the next. But it is only by seeing if they can breed and produce offspring (more seeds or pips, and more roses) that one can establish whether or not you're dealing with just one or two different species.

Darwin coined the phrase "artificial selection" to describe what human gardeners, farmers, and horse and dog breeders had been doing for centuries: creating better or more useful varieties of animals and plants.

George Washington did it as well. The father of my country spent considerable time, effort, and energy in the breeding of wheat. He used a magnifying glass and tweezers to get the pollen of one wheat stalk to fertilize the eggs or ova of another. Farmers continually control which stud horse gets the gig mating with which mare. If nothing else, they often make male horses into geldings. Ouch. Darwin observed that the processes of agricultural breeding are exactly the same process that takes place in nature. It's just that humans are exerting a great deal of top-down influence. Human farmers and breeders make the decisions as to which genes get selected to pass on to the next generation. Darwin thought of this as gene selection induced artificially (though he did not use the terminology of genetics, which lay far in the future).

On a philosophical note, we still use Darwin's descriptor or adjective: artificial. But notice that from the wheat's point of view, or the horse's point of view, or the point of view of the poodle, artificial selection is the same as natural selection. There you were minding your own business as a wheat plant. Your ova were fertilized by pollen. Wind may have brought them together. A bee may have visited a couple of other wheat flowers and shaken some outsider's pollen onto your eggs. Or a farmer might have carefully shaken pollen from one stalk onto the eggs of another. You, as the offspring plant, cannot tell the difference. From your point of view, having characteristics that appeal to another species is the same whether it's an insect pollinating you or a human.

It is interesting to note that people who believe in creationism, who hold that a deity set up all the ecosystems in just a few days, benefit in every way from our ability to breed better or more useful plants and animals. To get around this obvious contradiction they often resort to elaborate and arbitrary distinctions between what they do and do not call evolution.

Despite the underlying similarity, artificial selection is quite different from natural selection from an observer's point of view. In many cases, artificial selection serves human needs to the exclusion of natural fitness. A reasonable measure of this is whether or not an organism such as a wheat stalk, a soybean plant, a Quarter Horse, or a racing greyhound could or would survive without humans hovering over their affairs, seeing to it that they get fertilized or fed, and controlling their sexual interludes. For three out of those four, the answer is "no." Humans have made their lives possible. If you consider humans as separate from other living things (if you think a deity enabled and directed us humans to have dominion over Earth), then perhaps we are separate from nature and our genetic manipulations are our divine destiny. If, on the other hand, you see us as part of nature, part of a worldwide ecosystem, then our nominally artificial interference isn't artificial at all. We, and our activities, are part of nature.

Note that just as dogs are still dogs, most of these other domesticated strains still belong to their wild species. A Quarter Horse is still a horse, and spring wheat is still wheat. Artificial selection provides clues about how populations diverge enough to truly become separate species. For example, in the wild when populations of salmon get separated by a change in the flow of tributaries of the same river, the fish might look quite a bit different, one isolated population to the next. But the different isolated populations might still be capable of breeding, were they put back together in a salmon hatchery or separate eddy in the river. Some scientists refer to these as subspecies. If left apart long enough, or if enough generations are hatched separately, eventually we would expect the fish to become unable to breed across populations, and we would declare them different species. We expect that, because every time fish make new fish, every time they reproduce, there will be slight changes in their genes, in their DNA.

Eventually, we would expect too many changes for the subspecies to breed with one another. We should keep in mind, though, that just because we intuitively think of fish breeding as specific to one species and one species alone, it may be that nature allows a spectrum, with some able to interbreed and others not. They may look separate, like different breeds of dogs, while still being a part of the same species. The confusion between outward appearance and inner, genetic nature is also apparent—in a much more pernicious way—in the way people commonly talk about human races. (More on this in chapter 32.) Humans may have more trouble understanding nature than nature does is all I'm saying.

11

THE TREE OF LIFE—OR IS IT A BUSH?

People routinely use their "family tree" to describe a diagram of all of their relatives. It's such a common expression that you probably don't even think about the underlying metaphor when you say it. Tracing our ancestry lends itself to depicting one path of lineage. You and any siblings you may have all branch off from your mother and father. If you have children they branch off from you and your partner, and so on. It's a slightly confusing convention, since descendants should intuitively move down (descendants should descend, right?), but a tree grows upward (it ascends). Nevertheless, it's a very useful metaphor, not just for family relationships but for the far broader ones between different kinds of life. So let's climb on that metaphor and keep taking it further, step by step . . . or rather, branch by branch.

As you climb down from the uppermost tips of your family tree, where you are, to the limb below, and then to the limb below that, you are going to meet up with people who came before you—perhaps ancestors from your grandmother's spouse's side of the family. Farther down, you'll begin to cross paths with strangers; you'll meet people you never met, if you get my drift—ancestors so distant you've never even heard of them. Even farther down the family tree we will come upon the branch where *Homo sapiens* parted ways with other humanlike species. Now the tree begins to include big-picture evolution. If we keep climbing down toward the roots, we come across chimpanzees and orangutans. Below that we would find apes and our ancient common ancestor. Descending farther, we move past the primates and start running into other mammals, including lions, tigers, bears, narwhals, springboks, osprey, and bats.

If we continue our downward climb, we are going to come across branches leading to other organisms that, at first, might not seem that closely related to us (even my old boss . . .). I'm talking about lizards, fish, and marigolds. By logical extension, we realize that we must all be descendants of a single type of primordial organism at the very base of the tree. I know, I know; it's a little hard to believe at first. But at the microbiological level, we are all much more alike than we are different.

Every organism we can find here on Earth has DNA or its chemical partner, RNA. The acronym RNA is a contraction for ribonucleic acid, and it generally has just one strand compared to the more complex double-stranded structure that is DNA—deoxyribonucleic acid. ("Ribo" derives from an old word for sugar, which was made from a sweet compound called gum arabic. The adjectival form would be arabinose. From there, we end up with "ribo." Go figure.)

To me, the common code of life raises a powerful question, one that hits me personally. Through a surprising sequence of events in my life, I have become the chief executive officer of the Planetary Society, an organization cofounded by Carl Sagan, who was one of my professors in school. I took the job, because I cannot help but wonder if Earth is special or routine. And so I ask: Is the way life evolved here similar to the way life comes to be on other worlds, or are there drastically different evolutionary paths life can take? Put another way, is DNA, or something much like it, an essential feature of life? The only way to find out is—to find out.

But for certain people who hold a creationist point of view, life's common chemistry paints a completely different picture. They claim it indicates that we are all the product of a designer who made everything according to the same plan, all at once.

That line of reasoning also leads to questions—but they're the exasperating kind. If there was a designer, why did he or she or it create all those fossils of things that aren't living anymore? Why did the designer put all these chemical substitutions of radioactive elements in with nonradioactive elements? Why did a designer program in this continual change that we observe in the fossil record, if he or she assembled the whole system at once? In short, why mess around with all this messiness? If you're a creationist reading this, and you want to remark something like, "Well, that's the way he did it," I'll tell you right back, that is just not reasonable, nor is it satisfactory. If we were playing on a team right now, I'd say, "Get your head in the game."

Another thing: If there were a designer, I'd expect some better results. I'd expect no common cold viruses, for example. Or, if viruses are an unavoidable or accidental consequence of a designer designing with DNA molecules, I would hope that we'd be immune to those accidental viruses. If the argument is, "Well, that was all part of the plan," then I have to ask: How can you take the lack of evidence of a plan as evidence of a plan? *That* makes no sense.

Rather than ascending a tree, I like to think instead of a life moving along a time line, akin to my walk through time across the United States. This "Tree of Life" grows sideways, with time going from the distant past on the left to the present day on the right. Whatever we do to understand the branching pattern of a tree of life, we are working backward. We examine fossils and do our best to assess the age of the rocks in which the fossils are preserved. How else can we do it? We are constrained by the nature of time to start with where we are today and work our way farther and farther into the past, noting branches on a Tree of Life as we come upon them.

The way to peer into life's past is to examine fossils and to determine when a branch was first created by assessing the age of rocks when a certain fossil was formed. The oldest known fossils are of bacteria that apparently lived in ponds or a shallow sea located in what is now western Australia. They are fossil mats of bacteria, called stromatolites, whose metabolism led them to excrete calcium carbonate, chalk, the stuff of seashells and limestone. The ancient ponds dried up for some climatological reason, and these bacterial mats turned to stone. They're 3.5 billion years old.

As we examine rocks that are provably somewhat younger, we find evidence of more complex organisms. When we sample the floor of the ocean down just a few meters, we find a great many deposits of tiny sea creatures. These microfossils are beautiful, complex, cone- and mesh-shaped disks and spirals, which were the homes and bodies of ancient sea creatures. These tiny fossils are often the size of the head of a pin. If you know where to look, you can find hundreds of them in just a cubic centimeter of seafloor sediment. (A cubic centimeter is a milliliter in beer can and baby medicine measurement; it's the same as a cc in old-school medical terminology.)

The farther back in time we carry out an accounting of the number of different living things on Earth, the fewer different kinds we find. The implication is that evolution naturally leads to an increasing number of different kinds of living things. In this way, it is much like a tree. The higher a tree grows, the more branches will grow and bifurcate; each bifurcation leads to another branch and another bifurcation. That's why that Tree of Life metaphor is so powerful.

Darwin himself coined the phrase "tree of life," after he had drawn a sketch depicting the branching points in the natural history of living things. As we study the relationships between species today, we see a great many more species in the present day than in the distant past. This is true even after taking into account the five (or six) major mass extinction events. What all this diversity and creation of new species implies is that you and I, as animals on the Tree of Life, are related to all kinds of other organisms that we might not expect to be related to.

For a century and a half, scientists of all stripes (biologists, paleontologists, archeologists, pathologists, immunologists, astrobiologists) have been classifying all of life, living and extinct, to fill in all the branches on that Tree of Life. You might think, then, that by this time we'd have it all sketched out. Certainly you'd expect that we'd all agree about the main branches. Well, we don't. But we're workin' on it.

With the discoveries in the 1970s of organisms that were heretofore unknown to science, we've had to rethink which living thing is related to which other living thing. For a long time, scientists generally agreed that there were animals and plants. *Wow, Bill. Thanks.* Seriously, and they are still often given the organizational designation: Kingdom Animalia and Kingdom Plantae, as we still Latinize biology words. But after thinking about this in terms of evolution—that is, in terms of which organism came into existence before which other organism—scientists realized that animals and plants have much more in common with each other than they (or we) have with almost all of the other living things here on Earth. It's a stunning fact. Most of the living things on our planet are microscopic, and those microscopic organisms are less like you than you are like a cabbage.

Scientists have climbed down the Tree of Life; or you might say that they've moved right to left on the time line of evolution. Either way, they've reevaluated who seems to be related to whom. As I write, we now consider nature to have given rise to three or four foundational types of living things, or domains of life. I'm going with four. We have Bacteria, Archaea (microbes that are fundamentally different from bacteria), Eukarya (that's us, animals and plants together), and Vira. You could also call that last one Viruses. (I took some Latin in school, and I prefer this style of pluralization for this particular second declension noun, describing this particular domain of living or nearly living things.)

Not everyone agrees with me that Vira deserve a domain. A traditional argument is that viruses are not really, fully living things. They need a host cell to reproduce; they cannot reproduce on their own. Viruses make no effort to take in their own nutrients. They do not maintain a steady metabolism. They remain intact for extraordinary periods with no interaction with their surroundings. No energy comes in or goes out.

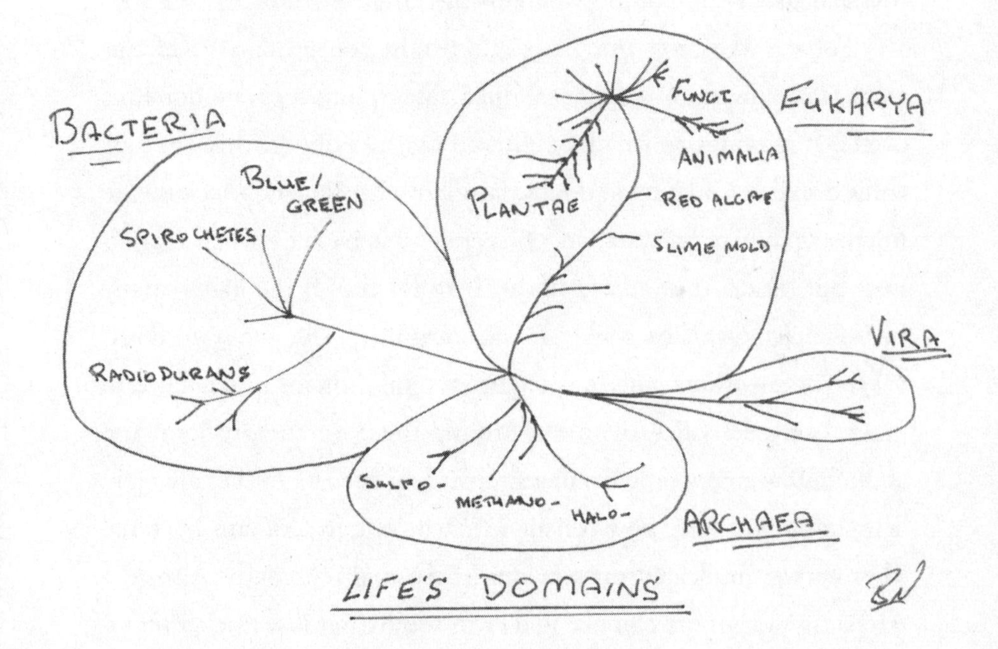

My view is that viruses would not exist at all if we did not have other domains of living things for them to parasitize. Vira interact with the other forms of life, which makes them more like us than they are different, and certainly more like life than like nonlife. The recent discovery of giant viruses like Mimivirus and Pandoravirus, so large that they blur the line between virus and bacterium, strengthens the case. At the same time, Vira clearly do not belong to any of the other three domains. For me, they should have their own significant branch on the Tree of Life.

After we've sorted out life's domains, it becomes a bit tangled as to whether we should continue to go with the old, increasingly fine designations. If we did, it would go domain, kingdom, phylum, class, order, family, genus, and species. O, would but it were that tidy. By diligently investigating the DNA and cellular structure of organisms, scientists have come up with terms like superfamily, subfamily, and infrakingdom. These words have been coined just to help investigators make sense of it all. I'll spare us the tangled details.

For me, the most important, and most confusing, part of this story is the division of Archaea and Bacteria into separate domains of life. It is a division that biologists recognized only within the past three decades. You are, no doubt, familiar with bacteria. Archaea are microscopic, like bacteria, and even resemble bacteria when magnified. But inside, they are significantly different. Of the many, many microscopic organisms, some have nuclei in their cells, and some don't. Of the ones without nuclei, some use several proteins in sequence to process or metabolize chemicals around them for energy. Some use about half as many proteins in a simpler sequence to do the same job. The simpler ones are now considered to belong to Domain Bacteria. The more complicated ones belong to Domain Archaea. Also, the Archaea have a more complicated or in some cases just thicker membrane of proteins and lipids around them than do the Bacteria. The Vira remain their own deal.

For people who study these domains, the difference is clear. Both the Bacteria and the Archaea do not have nuclei. Whereas, you and I do . . . in our cells. That's why we are said to be Eukaryotes; it's from Greek words meaning "having nut" (having nucleus). The Archaea and Bacteria are Prokaryotes, meaning "before nut" (before nucleus). This turns out to be of astonishing significance in the natural history of living things.

The next steps in classification become a bit blurry as scientists research and ponder the nature of our relationship to everybody else—that is to every other living thing here on Earth. If we continue to dabble in the term *kingdom*, one level of hierarchy below the domain, we get five or so. Everyone agrees that there are fungi. They get their own kingdom, the Fungi. Then there is another group called the protists that are often classified in Kingdom Protista (the malaria parasite belongs to this group). After that we have the single-celled Kingdom of Eubacteria. The eu- prefix designates the true bacteria. Then come prokaryotes—no nuclei inside. Then plants, the Plantae. And, the animals, like you and me, the Animalia.

In this way of looking at things, the least understood is the Kingdom of the Archaea. And yet they may be where life all started. It is an amazing insight.

If this version of the Tree of Life seems a little complicated, it is. But I must add: It's also just cool. Living things live and carry on with their business of living and reproducing whether or not you and I have a clue as to who gave rise to whom, or where the Vira fit in. But with the new genetic tools that let biologists assay the sequence of amino acids on DNA and RNA molecules, we are closer than ever to understanding which type of living thing came first. Then we can meaningfully ask a whole new set of questions.

Here's a big one: Why did all life apparently descend from just a single ancestor? Did some other type of primordial life take a shot at living and reproducing, but it just couldn't keep up? It's possible that life started more than once, and that you and I are the result of an ancient sorting out. I'll circle back to this idea near the end of the book in chapter 36.

12

BIODIVERSITY COMES IN THE TERRITORY

In recent years, scientists have devoted a great deal of effort to studying Earth's biodiversity, the total variety of life. Most often, people talk about biodiversity in terms of ecology and conservation, but there is much more to it. Biodiversity can be quantified. It is a measure of the results of evolution. It is like a master index of all the populations of all the species that have come into existence today and all those that have been lost to extinction.

When we look at the fossil record, we see that biodiversity has been generally on the rise since the beginning of life some 3.5 billion years ago: The Tree of Life keeps getting bushier. If indeed we all are descendants of a common ancestor, this is just what we would expect. With every reproduction, there is a chance for a mutation that may or may not prove to be beneficial to the offspring. If it's beneficial, that organism, along with its genes, survives well enough to reproduce, passing its genes forward one more generation. With this happening over and over and over, all day, all the time, around Earth, we end up with species after species distributed everywhere.

Confirming that trend was not a foregone conclusion. If the world and all its species of animals and plants were created at once by some supernatural force or event, we might expect nothing but the fossils of familiar, living species as we dig down in Earth's crust. Or if there was a before-and-after transition, as is described in *The Bible*, we might expect a great many fossils of now-extinct species in lower layers or strata of rocky formations, then a sudden break (corresponding to the end of Eden, or perhaps Noah's flood), followed by only modern species in more recent strata. But that is not what we observe, not even remotely.

There is no such division between pre-flood fossils and post-flood fossils, or pre-Eden and post-Eden ones. What we see instead is a continuum, one that shows the gradual trend toward increased biodiversity. That's exactly what Darwinian evolution predicts. Replication after replication, with chance after chance of small changes, leads to a gradual spread of good-enough traits that help a species compete. The opening of relatively unpopulated environments and the isolation of small populations encourage the emergence of new species. With more different kinds of living things coming into existence, more and different environments and energy resources could be exploited. Near as we can tell, what slows the natural increase in biodiversity is catastrophes like asteroids hitting Earth, smaller extinction events, like countywide mudslides, and, of course, us. Humans are apparently causing a mass extinction right now.

The evidence isn't flawless, of course. When we look at the fossils we have found all over the world and compile the fossil record, we have to expect limitations. After all, if a group of organisms was buried and eventually fossilized three billion years ago, we would expect fewer fossils to survive the trip through time. Tectonic plates shift and slide. Landscapes dry out and get flooded, often repeatedly. Any fossilized remains have a much greater chance of getting damaged, destroyed, or dissolved the longer they are buried in the earth.

Nevertheless even with that taken into account, we find that there is more diversity among the more recent fossils.

The driver of all this diversity is energy. In science education we like to say that energy is what makes things go, run, or happen. So it is with living systems. They (we) need energy to live, to move, to grow, and reproduce. If we ask ourselves, "Where is there the most energy available to living things?" we come up with two sources at least. The first is sunlight. The second is the primordial energy of Earth's interior. The same is probably true of life on other worlds (if it exists), as I'll discuss later.

Whether you're a green plant or something that eats green plants, or something that eats things that eat green plants, you are going to find the most sunlight near the equator. No surprise then that there are more living things in a tropical rain forest and around tropical coral reefs than there are on the open ice of the Arctic and Antarctic, human scientists stationed there notwithstanding. But along with the multitude of individual living things, there are also more different kinds of living things near the equator than we observe in the north and south polar regions.

Furthermore, there is a gradient of diversity. There are more different kinds of living things per square meter or hectare or square yard in the Amazonian rain forest than in the rain forest of Belize or Guatemala. They each, in turn, have more diversity than we find in the boreal forests of Northern Canada. There is a little more diversity on New Zealand's North Island, which is closer to the equator than New Zealand's South Island, which is closer to the South Pole. There are other local factors, rainfall especially, but the trend holds in the big picture. If you think about it, you may have observed evidence of this phenomenon yourself. If you live in or visit the U.S., compare the thick growth in the wetlands of Louisiana with the headwaters of the Mississippi River in Minnesota.

The areas of Earth with the most energy input are also the areas with the most biodiversity. This gradient from more to less as you go north or south from the equator is yet further evidence of evolution. These ecosystems have been there for a long, long time, and the longer an ecosystem is running, the more living things can multiply. As they increase in number, they'll carry more mutations and more variants. With diversity in offspring and lots and lots of time, we end up with a biodiverse ecosystem.

The patterns of energy are very different for living things that rely on Earth's internal energy. Nuclear fission of natural radioactive elements like uranium and thorium keeps the insides of the planet molten. We experience that energy directly when that heat finds its way up close to the surface. It drives steam geysers, deep-ocean vents, volcanoes, and earthquakes. But over the past few decades, scientists have discovered that it also drives whole ecosystems at the bottom of the ocean, a previously unrecognized realm of biodiversity.

Exploring the deep ocean is difficult. It's cold, it's corrosive, and it's crushing (the sea's three Cs). Any equipment we send down has to be able to survive these rigors. On top (or bottom) of that, it's dark. The illumination from lights aboard deep-ocean submarines or submersibles is quickly absorbed as it passes through just a few meters of seawater, so the images we can return are just perhaps the size of a large living room. As we study the landscape of the deep-ocean floor, we see only the occasional sign of life—except at deep-ocean vents. In these extraordinary locations, geothermal heat provides the energy for ecosystems to thrive in the dark. There are enormous red-tipped tubeworms, strange fish, albino crabs, and clams the size of footballs. When brought to the surface and opened, they look like a steak but smell like a swamp.

The animals living around deep-sea vents are different than those near the surface. Their metabolism is based on chemical exchange with the hot, nutrient-rich seawater. We call the process chemosynthesis rather than green-plant photosynthesis. Clams down here need heat and hydrogen sulfide (poisonous to you and me), whereas clams near the sea surface depend on photosynthesis of plankton that they siphon through their digestive systems. As fascinating and just plain weird as the deep-sea geothermal vent ecosystems are, they have a great deal less diversity than we find in ecosystems that receive direct sunlight. At deep-sea vents we've counted about 1,300 species so far. In the Amazon rain forest, we can find 40,000 species of insects, just insects, in a typical square kilometer. Couple that with trees, monkeys, spiders, and snakes, and the rain forest has thousandfold the diversity. Why would that be?

Fundamentally, there is just less energy to be exploited in the very deep sea. Some of these hydrothermal vents run at 400° C, but that hot spot is concentrated in a small footprint—just a few hundred known locations along volcanically active strips on the ocean floor. (The water at this depth does not boil because it cannot form vapor bubbles under the pressure of the weight of the ocean above.) In contrast, solar energy falls on every part of the planet's surface, with an intensity of up to 1,000 watts per square meter.

I pursued this digression about the deep-ocean vents because it provides further insight into how evolution works. Fewer species per square meter in the deep ocean than in the brightly lit forest is exactly what we'd expect. Up here, where we live, there is more energy to drive more living things. They reproduce more quickly and we end up with more diversity. Deep in the cold ocean, life thrives only where there's enough energy to feed the system. There is not nearly as much energy down there to run the biodiversity machine.

In the 1990s when I was doing the *Bill Nye the Science Guy* show, we did a whole episode on biodiversity (show number 9). At that time, we were pretty sure that the most diverse ecosystems of all are not in the middle of one of the world's 292 major river systems. They are not in the shallow sea, at a coral reef, say. They are probably in between—in estuaries, where rivers meet the sea. Since then, it's been suggested that rain forests near the equator are the most diverse ecosystems. In either case, we find the greatest diversity where there is a great deal of freshwater.

When you look at a picture of Earth from space, the ocean is the biggest contiguous area you can see. So at first, you might assume that the ocean would be the place to find the most of everything, the most biodiversity. Furthermore since most of us, that is most living things, are full of liquid water, we can figure that life started in the ocean, and with all the extra time that ocean life has had to evolve, you'd expect results. You might think that any place in the sea where there's sunlight and enough mixing of deep-water nutrients would be the place of greatest diversity. But generally speaking, the ocean is not where we find the most diverse ecosystems.

To be sure, coral reefs carry enormous diversity. Having dived in coral reefs in Hawaii, the Pacific Northwest, and along the California coast, I can tell you that there are more different kinds of fish here than you can name in an hour. I often reflect on the species I can't see: the bacteria, viruses, transparent cnidarians (sea jellies, once commonly known as jellyfish), and rock-look-alike porifera (sponges). There are thousands upon thousands of species within just a few flipper strokes of anywhere you swim.

In the same way, when I went walking through the rain forests along the Sibun River in Belize, the Whirinaki River in New Zealand, and the Hoh River in the United States, I couldn't get over how much was going on all around me. When you walk in any of these places, you sense that there are countless species swarming, sprouting, hunting, and being hunted all around you. If you told me these are the most biodiverse places on Earth, I'd believe you.

From your own experience, you may know that you cannot drink salt water. It just makes you sick. You may also know that, except for a few remarkable species, you can't put a saltwater fish in freshwater or the other way around. The fish will die. You may have run a classic experiment in school demonstrating what chemists call osmosis. If you dissolve the shells off of uncooked eggs using vinegar, then place an intact exposed egg in distilled water and another in salt water, you'll observe the water molecules slowly passing through the membranes to the saltier environment. The egg in distilled water will swell. The egg in salt water will shrink. This same membrane chemistry would, one might think, keep the two types of ecosystems separate. To a degree, they do remain apart—except at the estuaries.

In estuaries, where river systems meet the sea, there's a mix of freshwater diversity with ocean-water diversity. Instead of either system winning out or cancelling the other type of system, they work together. The likely reason for this is that ecosystems with a lot of diversity can adjust as the environment changes around them. This is another testable evolutionary idea, and scientists have set out to see if it is true—if diverse ecosystems really are more robust overall.

It's possible to quantify the resilience of an ecosystem by measuring the number and mass of all the living things before and after a big change in environmental conditions. If there's a drought, or an unusual amount of rainfall, or sudden temperature swings to freezing or sweltering, the more diversity a system has, the better its species do at staying alive and reproducing. That's the hypothesis. There are at least two ways to investigate it. We can study systems where biodiversity has been reduced and where it has been increased, or we can make the same observations on ecosystems at each end of a spectrum. With either, the theory holds up. More diversity makes for a more robust ecosystem. That explains why the estuaries are so rich. Both the freshwater and saltwater ecosystems there are already full of species. The thick web of life helps saltwater organisms to adapt to some freshwater, and freshwater organisms to adapt to some salt water. Diversity begets more diversity.

The opposite also seems to be true: Places with low diversity are at a greater risk of losing even more. Finding places where the diversity has been decreased is nearly effortless, unfortunately. Humans have wrought havoc on so many environments worldwide that there is almost no such thing as an unspoiled natural area. I climbed mountains in the Pacific Northwest for many years and remember well reaching the summit of Mount St. Helens and looking north to the magnificent vista of Mount Rainier and seeing a distinct layer of smog. You didn't have to know anything about air pollution to conclude that the smog flowed in there from our beloved cities of Seattle and Portland. Sure enough, airborne insect populations are affected. The smog cuts into their numbers a little bit. In turn, the number of plants that can start growing in the volcanic soils below is reduced a little, because the insect carcasses provide less nitrogen for plants to take hold. The view is still gorgeous, but the scene is not quite pristine.

If you really want to get the creeps, visit a Confined Animal Feeding Operation (a CAFO). Wow. Livestock are kept confined and fed a diet to fatten them fast. Their hooves destroy any grazing area. Their excrement poisons everything it flows through and to. And of course, these cattle are fed a surfeit of antibiotics to suppress the diseases that can easily be communicated from one animal to the next, which in turn leads to the rapid evolution of those disease parasites, which in next turn renders those same antibiotics ineffective. And on it goes. While we're eating meat, we're producing new strains of diseases and destroying watersheds. I'm pretty sure that by understanding this process, we can do better. Here's hoping.

On modern large-scale farms, we see thousands of hectares or acres of a single crop. While this "monoculture" makes it easier for farmers and farm machinery to harvest the crop when it's ripe or ready, it also makes the crop more susceptible to attack by a single pest or parasite. If you are a corn-borer moth, you and your swarm can have a literal field day eating every cornstalk and ear of millions of corn plants across many thousands of acres. It's true for human-built farming systems, and it seems to be true in nature. When a monoculture gets established, it's vulnerable to problems.

Often forest ecosystems—especially temperate forest systems—look monocultural. From the air, western Canada looks like a limitless expanse of fir trees, for instance. But there is an important and not-so-subtle difference between a large natural stand of evergreens and a human-planted tree plantation: age. In natural stands there are trees of all ages. Old ones are tall; young ones are short. More important, the fallen trees provide nutrients for the next generation. There are microbial systems in the forest floor that support the root systems of living trees that are photosynthesizing and growing. There is a tremendous amount of invisible biodiversity in that seemingly unified evergreen forest.

In the Pacific Northwest in the springtime, the pollen falls so thickly that it looks like a yellow haze of fog. Even if you consider yourself allergy free, you can feel the pollen in your sinuses. It appears monocultural in species, but it is not so in age and stage of decay. It is hardly uncommon to hike through the Alpine Lakes Wilderness and see young tree after young tree literally growing right out of a fallen or felled old tree. By charming tradition the old dead tree's base is called a nurse stump. It's nurturing its progeny. In the cases where humans did the deadly sawing, "nurse stump" is an ironic turn of phrase.

Scientists have run remarkable tests showing the effect of biodiversity. In bare fields, researchers have planted essentially monocultures of grasses in one area and in another field an assortment of grasses tenfold more diverse in species. In the first few years, the monoculture field shows a lot of growth. It looks, at first, like the monoculture creates more plant matter, more so-called "biomass" than the diverse fields. But over the course of a few years, a decade or so, the field with the diversity of grass species wins out, producing more biomass and more healthy plants—and as a result a healthier group of animals that live off the grasses—than the monocultural cultivation.

This is apparently because diversity leads to an improved ability of an ecosystem to absorb changes in things like weather, climate, or the introduction of some new species. Try this as a simplified example. Let's say we have a monoculture meadow and a diverse meadow. In a diverse meadow, the different species of grasses and flowers produce pollen at slightly different times of the year, varying their seed and pollen production by a few weeks or even a few days. Pollinators like bats, birds, and bees can find steady work. They'll be there for each different pollen production cycle. On the other hand, in a meadow with a monoculture, where all the grass pollen is being produced at once, there is likely to be too few pollinators around. The bat, bee, and bird populations cannot sustain themselves between the cycles of abundance of the nectar and pollen. The grasses, the birds, the bees, and the bats all suffer a little. Without diversity, each of the species is less successful. Diversity leads to resilience.

You see the impact of humans on Earth's environment every day. We are trashing the place: There is plastic along our highways, the smell of a landfill, the carbonic acid (formed when carbon dioxide is dissolved in water) bleaching of coral reefs, the desertification of enormous areas of China and Africa (readily seen in satellite images), and a huge patch of plastic garbage in the Pacific Ocean. All of these are direct evidence of our effect on our world. We are killing off species at the rate of about one per day. It is estimated that humans are driving species to extinction at least a thousand times faster than the otherwise natural rate.

Many people naïvely (and some, perhaps, deceptively) argue that loss of species is not that important. After all, we can see in the fossil record that about 99 percent of all the different kinds of living things that have ever lived here are gone forever, and we're doing just fine today. What's the big deal if we, as part of the ecosystem, kill off a great many more species of living things? We'll just kill what we don't need or notice.

The problem with that idea is that although we can, in a sense, know what will become or what became of an individual species, we cannot be sure of what will happen to that species' native ecosystem. We cannot predict the behavior of the whole, complex, connected system. We cannot know what will go wrong or right. However, we can be absolutely certain that by reducing or destroying biodiversity, our world will be less able to adapt. Our farms will be less productive, our water less clean, and our landscape more barren. We will have fewer genetic resources to draw on for medicines, for industrial processes, for future crops.

Biodiversity is a result of the process of evolution, and it is also a safety net that helps keep that process going. In order to pass our own genes into the future and enable our offspring to live long and prosper, we must reverse the current trend and preserve as much biodiversity as possible. If we don't, we will sooner or later join the fossil record of extinction.

13

FOSSIL RECORDS AND EXPLOSIONS

When people talk about ancient life, they frequently talk about the "fossil record." But let's be clear: The fossil record isn't a tidy clean recording. No one went to a studio and methodically laid down some tracks, the way a rock and roll band records an album. The evidence embedded in Earth's rocks is more like the work of a band that recorded with a faulty microphone and then accidentally recorded over most of the tracks. On top of that, when they were finished, they lost almost all of the final versions. Most living things never get fossilized, and most fossils end up in places where they are impossible to discover. It is impressive that paleontologists have been able to reconstruct the lives of individual plants and animals. It is downright amazing that they have found enough fossils to deduce the broad sweep of evolution's great busts and booms—including one wild episode of innovation that led to all of the large animals living today.

Let's step back for a moment and consider what it takes to become a fossil. First of all, you have to get buried. I don't know about you, but I'd prefer to be dead first. However, from a fossil hunter's point of view, the less time the deceased has been left out in the sun, or washed around in the sea, or scavenged upon by chewy-beaked birds, the better. Creepy as it sounds, getting your specimens buried alive is the best thing for a paleontologist. The resting place generally has to be wet in order for the organism to get buried effectively. Then that wet sand or soil has to dry out completely so that microorganisms don't rot the remains. Then it has to sit there for years and years and years, generally millions of years, while minerals slowly trickle through and turn once-living structures to stone.

This extraordinary string of events produces about the most incomplete kind of record keeping you can imagine. Almost every animal and plant that has ever lived has disappeared without a trace. Suppose a fossil were perfectly preserved, but then got subducted (pulled downward) under one of Earth's continent-sized tectonic plates. It would become part of the magmatic mantle. It would melt into molten rock. The pattern would disappear like a corporate banquet ice sculpture on the next day's afternoon.

It is only through very unusual happenstance that we are able to find any trace of the past. That's why, when exciting new (or are they old?) fossils are found, it is—thrilling. Notice also, it's probably easier to find big fossils. A tyrannosaur femur is easier to come across than a pigeon-sized animal's toe, even if they're both preserved in the same fossil bed. This selection effect may have influenced our understanding of ancient dinosaur ecosystems. We might presume the big creatures dominated, when perhaps actually there were a great many smaller animals that just don't show up as often at our dinosaur digs.

The deeper we dig, the older the animals and plants we find. We can trace the development of certain species for millions of years. Trilobite fossils are so plentiful that scientists are able to classify them into orders, suborders, families, subfamilies, and so on, all the way down to genus and species, just like biologists do for living creatures.

We can trace the trilobite lineage back more than 250 million years. Trilobites remind me of crabs and lobsters. They had hard shells that held their shape as they were buried and slowly turned to stone. When I went on walks in central New York State, I would find trilobite fossils all over the place. On the other hand, there are relatively few fossils on Earth of animals with soft body parts. When we find fossils of ancient rhinoceroses for example, we almost never find fossils of their ears. Soft things tend to decay instead of fossilizing.

There are a few important examples of fossils that include the soft and hard body parts of long extinct creatures. They figure importantly in the history of fossil finding and the history of our understanding of evolution. I refer to some fossils found in shale, the sedimentary rock that old-style classroom blackboards were made of. The shale found in the Burgess Shale Formation in Canada is the most well-known, because the fossils are just so perfect and because it covers a truly extraordinary period in Earth's history.

In 2005, when I was doing a television show about great scientific discoveries, I had the wonderful opportunity to hold a few Burgess fossils in my hands. These fossils are astonishing, with silvery traces and lines preserved in layers of very smooth almost black rock. This solid rock can be broken apart pretty easily in perfect, flat, distinct, obvious layers. Skilled rock hounds can gently tap open pieces of the shale with their ever-present rock hammers, so that the layers split apart like pages in a book. The Burgess area is in the Walcott Quarry in British Columbia. It's all part of the Canadian Rockies now, but it was once a great big wall of mud and part of an ancient ocean reef, about 500 million years ago, during the Cambrian Period, which is named for an area in Wales, in the United Kingdom. (By long tradition, geologic time spans get their names from the place in which they were first discovered or cataloged. It's coincidental that both Wales and the Burgess area were once part of the British Empire.)

By looking things over carefully, paleontologists have surmised that an ancient huge wall of mud gave way all at once. It slumped or slid downhill extremely quickly and buried countless sea creatures in a slurry of fine silt and seawater in just a few moments. There is no evidence in the Burgess Shale Formation of any of the creatures struggling to get free or burrow their way out. They must have been immobilized and suffocated in a snap. If you think too hard about those moments, it can be troubling. But for us learning about our place in the process of evolution and the progress of life here on Earth, it is an extraordinary find. In no other excavation site ever explored have the fossils been of as high a quality as those found at the Burgess Shale Formation, apparently because the ancient silt was so fine-grained.

The Burgess Shale Formation is full of beautifully preserved shells and flesh from dozens of animals that were absolutely unknown to science before the shale's discovery by paleontologist Charles Wolcott in 1909. Even then, the age and significance of those fossils was not appreciated until 1966, when a number of researchers reexamined the site and recognized the true age and diversity of the creatures preserved there. By carefully photographing, carefully tracing, and then carefully abrading the rocky slate prisons of the images, fine layer after fine layer, investigators were able to reconstruct the sizes and shapes of ancient creatures. They realized that they were looking at animals that swam in weird ways, walked in weird ways, and captured prey in ways that many never thought possible. What animal do you know that has five eyes, for example—two pair and one in the middle—as Opabinia once did? His and her brains would have to process images in some way that we even-count-eyed folk can't quite imagine. How about an animal so confusing to the world's experts that it was given the name *Hallucigenia*, because they felt like they were seeing things.

Some evolutionary biologists have argued that the abundance of fossils preserved in shale from this part of the Cambrian Period, about

500 million years ago, indicate that life on Earth was once much more diverse than it is today, but most disagree. The quite reasonable presumption is that these crazy-looking creatures arose by natural selection, but their body plans proved to be unworkable in the long run. They got outcompeted by the more recent animals and their ancestors that we observe today. These crazy critters then disappeared from subsequent layers in the fossil record.

As I consider these animals, they don't look—to me—that much different from what we see today, if one accepts the idea that each appendage and each organ served those animals, and in modified forms they serve the invertebrate sea creatures of today. Then again, you can look at these creatures and focus on the specific ways in which they are drastically different from anything swimming around now. They are from an ancient, completely unfamiliar underwater world.

The Burgess fossils are unique in their preservation of life in the Cambrian, but to me it is not so extraordinary that this site is so extraordinary. It is difficult to produce fossils in nature in the first place, and extremely difficult to find fossils of soft body tissue anywhere for any geologic period, period. It seems reasonable to me that if we just knew where to look, we could find other extraordinary animals and plants preserved in the fossil record. After all, more than two thirds of our world is under water. I imagine that there are countless rich fossil beds that we'll never be aware of because they're in deep water and buried in modern or much more recent detritus; that's the fine sediment that falls from the surface as plankton live and die above. The Burgess Shale Formation was discovered a full fifty years after Darwin published his seminal work. I have to think that there are plenty more places to look.

Along with the stunning fossils themselves, the Burgess Shale Formation is notable for the era of life that it records—a period of breakneck evolutionary innovation called the "Cambrian Explosion." Over the course of 20 million years, the number of novel species in the fossils found around the world increased twentyfold. This burst of diversification is often presented as a big evolutionary mystery. Creationists, in particular, often talk about the Cambrian Explosion as if it happened in an instant. This is, for me, another example of extraordinary ignorance or very limited critical thinking.

For starters, I have to quibble with the word *explosion*. I've been at rock quarries, when the engineers and technicians are, well . . . blowing stuff up. A typical shot, or explosion sequence, involves a few thousand charges and runs less than thirty seconds. When a sequence takes millions upon millions of years, should we call it an explosion? In geologic time, 20 million years is not very much, but it's not like the milliseconds between explosions in a rock quarry shot. It's a long time.

Along with that, is it that life on Earth actually increased in diversity by a factor of twenty, or is it that the fossil record doesn't reveal much about the past until living things came up with hard shells that were readily preserved in solid rock as fossils? The Cambrian Explosion is more (or was more) like inflating an enormous air mattress. Once some marine organisms evolved hard shells, it's not surprising that these creatures were successful and diversified. If you were to puff on almost anything for 20 million years, you might expect an increase in volume, no?

For me, the idea that there used to be more diversity than we see today is reasonable, but not likely. There have been mass extinction events along the way (more on that in the next chapter), but they don't seem to be, if I may say, *massive enough* to drive us to this conclusion. In the same vein, the Cambrian Explosion strikes me as more of an artifact of the fossils we can find in rocky layers corresponding to about twenty million years rather than an actual very, very fast production of diversity in species. Instead, there was an increase in the size and robustness of the shells of invertebrate sea creatures that are inherently better preserved as fossils than were the animals and plants that preceded them. But of course, I may be wrong. Please investigate this business for yourself and reach your own conclusion.

In general, the difficulty of making fossils guarantees that the fossil record will be full of gaps (or maybe "skips," for those of us youthful deejays or old fogies familiar with vinyl records). It could not be any other way. The smaller the organism, the softer its body, the longer ago it lived, and the more likely it is to have fallen into one of those gaps. The amazing thing is not that we do not have more examples like the Burgess Shale Formation; the amazing thing is that we have even one near-perfect specimen.

More secrets are out there, waiting to be brought out into the open. Investigators working with existing collections and rock hounds climbing on cliffs today are continually expanding the fossil record. As we move into the future, we keep learning more and more about the past. Every day we each have less life left to us, and yet every day we have access to a greater and greater stretch of evolutionary time. I frankly hope that one of you (or me) reading this work finds a fossil—or becomes a fossil someday—so that the record for our descendants will be that much more complete.

14

MASS EXTINCTIONS AND YOU

For my debate with creationist Ken Ham in Kentucky, I estimated that there are about 16 million species of life on Earth. What if 14 million of them suddenly disappeared? Sounds hard to believe, like some dystopian Hollywood summer blockbuster? But that is exactly what seems to have happened at least five times over the past half billion years. In each instance, a catastrophic event (or combination of events) killed up to 90 percent of the world's sea and land creatures, and it happened in the blink of an eye—geologically speaking, at least. Other than the origin of life itself, mass extinctions are the most dramatic and mysterious events in the history of life on Earth.

Some of the best evidence for these mass extinctions is found in the rocks that make up, or once made up, the ocean floor. Scientists head out to sea with massive but elegant devices to cut cylindrical samples out of the ocean floor, where sediments are laid down with some regularity. And once down there, the sediments don't get weathered very much. There are no winds, or rains, or freeze-thaw cycles. The Earth is 4.54 billion years old. Its crust has repeatedly broken up, moved around, melted, and reformed. Most of our planet's extraordinarily ancient history is lost to us. Yet locked away in ancient sediments, some evidence of those five extinction events still remains.

One of the great challenges in reconstructing a mass extinction is making sense of what happened when. In the same way we have divided living things into a hierarchy of divisions—domain, kingdom, phylum, class, order, family, genus, and species—geologists have broken apart the long history of our planet into eon, era, period, epoch, and age. As biologists understood the relationships between living things more completely, they added other terms, like superorder and subspecies. In similar fashion, geologists sometimes use the term *supereon* to describe all of Earth's history before the Cambrian, from the planet's formation through the Precambrian.

You have probably heard terms like Mesozoic Era, Jurassic Period, and the Stone Age. These phrases come from geology. For seven eighths of Earth's history—basically, for most of it—living things were just revving up. Most life was single-celled organisms or relatively simple, soft-bodied animals. The last 500 million years are when almost all of the interesting stuff happened. Pretty much every creature you have ever heard of, from trilobites to dinosaurs to Neanderthals, appeared during that time. All of those years are currently considered to be part of just one single geologic eon, which goes by the wonderful name: the Phanerozoic Eon (Greek for the visible eon, the one we can see). The Phanerozoic is in turn divided into three geologic eras: Paleozoic, Mesozoic, and Cenozoic. That last one is the one we're living in. The names mean, roughly, old animals, middle animals, and new animals (which includes us). Finally, those eras are divided into epochs. Ours is the Holocene or "recent" Epoch, covering just the last 10,000 years—about 1/500,000th of the whole history of the planet. I drew a sketch.

All of those names and dates are essential to the paleontologists who are analyzing the history of evolution. They have painstakingly analyzed the available fossil evidence and assessed the number of

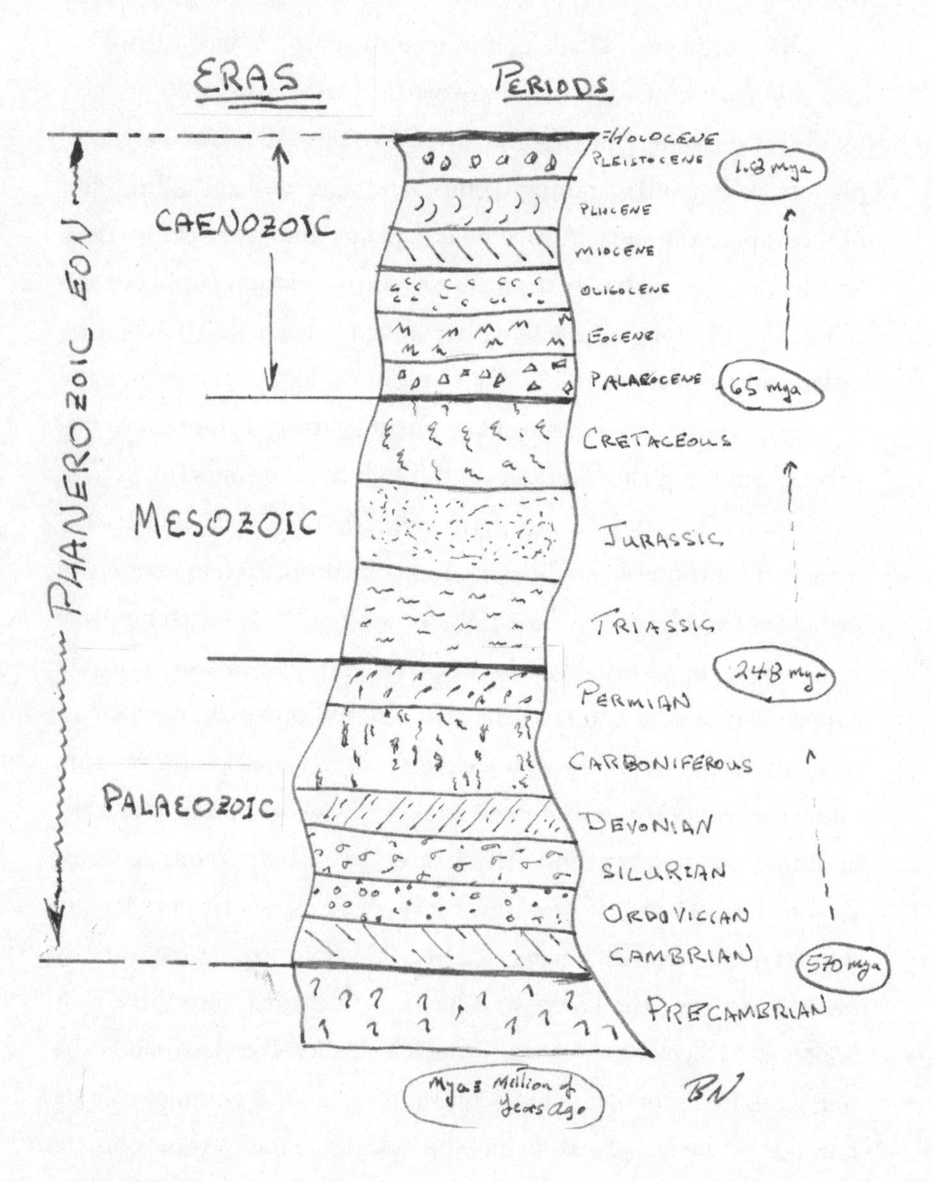

living things on Earth, especially in oceans, at different times over the last 500 million years. In the process, they've discovered relatively short intervals of time when the amount of diversity has dropped abruptly. Those are the mass extinctions. I'll talk here about the five we know about. But I hope we'll all soon acknowledge that there really is a sixth mass extinction, one that is happening right now.

The oldest of the mass extinctions is called the Ordovician-Silurian, from the names of the geologic periods when it occurred (444 million years ago). Here we lost about 85 percent of ocean species. At the time, there was essentially no life on land. The ocean was where all the action was. It was the second greatest of the mass extinctions. Next we had the Late Devonian extinction, about 364 million years ago. This time we lost about half of Earth's animals and plants.

The most traumatic episode in life's Phanerozoic history was the Permian-Triassic extinction, which went down 251 million years ago. As a young engineer, I worked in the Permian Basin in Texas. I had no trouble observing the ancient seashells on the ground, right next to the oil-rig plumbing I was servicing. It's the same Permian you may have heard about if you enjoyed the television show *Friday Night Lights*, about the drama that is west Texas high school football. Near as we can tell, Earth lost about 75 percent of all species during the Permian-Triassic extinction. Think of it! At the end, only 25 percent of everything was still alive. You and I are descendants of that lucky minority, the 25 percent.

Where I went to college in central New York State, it's not so difficult to walk in the beautiful shale gorges cut by streams and find a trilobite fossil. In a week, you can find a dozen. It is sobering and remarkable to realize that trilobites lived here for more than 250 million years, from the Cambrian right up through the Permian. Yet they are all, every one of them, gone, extinct. I'd rather that didn't happen to us, at least not for a few more periods. On the other hand, the Permian-Triassic extinction cleared the way for the rise of the dinosaurs. It shows how robust living things are, and how strongly living things can rebound . . . given enough time, time, time.

Next we have the End-of-the-Triassic extinction, about 200 million years ago. During this period, we seem to have lost about half of the world's species. By this point in the story, it almost sounds routine.

Then at last, we arrive at the most well-known and significant mass extinction, the Cretaceous-Tertiary extinction, 66 million years ago. Geologists like to use single letters to designate the geologic periods. Who wouldn't? It saves a mouthful. As we move from the deep past into the more recent past, we already have two periods that begin with the letter C, the Carboniferous and Cambrian, which are written as C and Ꞓ. So the Cretaceous came to be abbreviated as K, and the die-off between the Cretaceous and Tertiary came to be called the K-T extinction. The K has a legitimate pedigree: It comes from *Kreide*, the German word for chalk, which is where Cretaceous got its name in the first place. More recently, geologists have given the early part of the Tertiary a more specific name—the Paleogene. So here in the early twenty-first century, we mark the Cretaceous-Paleogene boundary in the geologic record. The mass extinction that included the ancient dinosaurs is called, in abbreviated form, the K-Pg extinction.

By its old or new designation, this extinction is the dramatic one that marked the end of the age of dinosaurs (although I like to remind people that we are still surrounded by the modern descendants of the feathered dinosaurs—the birds). It's famous for most of us who are young at heart and fascinated with dinosaurs. It also, not incidentally, ushered in the age of mammals. It made space for us.

We cannot be absolutely certain what caused any of these mass extinctions, but we have a great many excellent clues. Just as important, we have a great many mathematical models of Earth's climate. We do our best to estimate what it would take to bring about traumas of these magnitudes. We look at the rocks, the fossils, and the chemistry. We run the numbers and come up with very thoughtful hypotheses about what happened. Studying the past is very helpful; it enables us to make better predictions about what could happen on Earth again, perhaps very soon.

We also have much more direct access to the here and now. Looking at Earth from space, as we can nowadays with our sophisticated satellites, we can observe climate as it is changing today. Not only that, but we can compare the climate of Earth with climates of our nearby planetary neighbors, Mars and Venus. We can infer just what it might take to change the climate of a whole planet so radically, and in such a short span of time, that half or even almost all of the living things there die and disappear.

We collect additional clues by studying Earth as a complex living system, a planet-wide ecosystem. If you consider any ecosystem that you've ever lived in—a forest, a city, a farm, or perhaps you've sailed on the ocean for a time—you can see these systems are complicated. Living things interact with their environments in countless ways. When the environment changes quickly, ecosystems change as well. When I worked for Boeing, I spent many wonderful hours hiking and climbing in the Seattle area. In the mountainous western part of North America, you can walk right up and over areas where rockslides have occurred in the last few hundred years. You can see the radical differences in plant life and wildlife between the top of the slide, the bottom or toe of the slide, and the unchanged boundaries, where the trees and fauna are living pretty much as they did before falling rocks crushed and scraped a great many living things down the mountain.

In the case of mass extinctions, it must be like a rockslide on a global scale. What could bring about something like that? For me, there are two things I can imagine right away. Geologists have, too.

The first potential extinction triggers are volcanoes. If you ever get a chance, I strongly encourage you to visit Mount St. Helens National Volcanic Monument in Washington State in the U.S., where entire ecosystems vanished when the mountain blew its top on May 18, 1980. Countless birds, fish, insects, and hundreds of large animals like deer, voles, and raccoons were killed in an instant. All evidence of them was either buried under thousands of tons of massive icy, rocky, mud, or it was incinerated, burned to a crisp. If you can, go to Hawaii to see the state's volcanoes oozing red-hot molten rock. There's no stopping a lava flow; it incinerates everything in its path. Imagine dozens or hundreds of volcanoes spewing ash and fire over enormous areas of the planet's surface. The result could be such an abrupt change in Earth's atmosphere that no living thing, especially no complex system of living things, would be able to stay alive.

We know that mass volcanic eruptions occurred because we can still see the giant lava flows that resulted. These bursts of volcanism are comprised of the distinctive rock called basalt. When it cools, it often forms enormous blocks with right angles like gigantic grains of table salt. The lava flows are so expansive that geologists call them flood basalts. Some of these enormous flows could have erupted even under the sea surface—such as the giant Kerguelen Plateau in the south Indian Ocean—radically changing the oceanic and atmospheric chemistry. An enormous outpouring of lava in what is now Siberia is currently considered the best explanation for the devastating Permian-Triassic extinction. Researchers are blaming volcanoes for the end-Triassic extinction, too. But keep in mind: We're investigating an old, old crime scene.

In the Deccan region of what is now India, there is an enormous zone of volcanic rocks. The zone is bounded by India's east and west coasts and the Vindhya Mountains. Outcroppings of these layers of rock resemble stair steps, and the Scandinavian word for stairs is *trapp*, which when shortened is trap, so geologists embraced it. At the Deccan Traps we find several layers of rock that cover an area of 500,000 square meters (200,000 square miles). They comprise 1.2 cubic kilometers of lava. It was one hell of an eruption or series of eruptions.

In between the layers of frozen lava in the Deccan Traps are layers of sediment laid down by ancient seas. Geological dating of those layers shows that they formed between 60 and 68 million years ago and that the eruption(s) reached a peak around 65 million years ago—about the same time the ancient dinosaurs met their fate.

Could these flood basalts have something to do with the demise of the ancient dinosaurs? Some geologists, like Princeton University's Gerta Keller, have made a strong case. After some fieldwork in the area, she said, "It's the first time we can directly link the main phase of the Deccan Traps to the mass extinction." She was talking about the ancient dinosaurs, et al.

When Keller studied the fossils in ancient sediments near the Deccan Traps, she saw that the biodiversity of foraminifera (a broad class of aquatic microorganisms) took a tumble right around the time of the eruptions. Apparently, there were at least significant local extinctions. The powerful volcanoes that built the Deccan Traps must have spewed toxic gasses and created enormous layers of atmospheric dust; those emissions reflected sunlight into space cooling Earth somewhat. During other episodes, volcanoes shot out tons of greenhouse gasses, heating the world up in a hurry. It was climate change with whiplash.

Even a less violent form of geologic change could have been devastating to life. For instance, the shifting of the continents and shorelines could have sent the global climate into an inhospitable new state. That seems to have happened in the Ordovician-Silurian, when most of the land on Earth was part of a single supercontinent that migrated to the South Pole. During this period Earth cooled, enormous glaciers formed, sea levels plummeted, and a lot of ocean life was left high and dry. But some extinctions seem to have happened quickly, not just in geologic terms but in human terms as well.

Which brings me to the second big extinction-driver: asteroids. If Earth is struck by an asteroid, everything can change in a flash. Such an event may have brought on the K-Pg extinction, when things seem to have changed far too quickly to explain with a volcanic eruption—at least, not with a volcanic eruption alone. The scientific consensus today is that the main blow to the ancient dinosaurs was a 10-kilometer-wide rock that struck our planet off the coast of what is now Mexico. The result is still faintly visible as a 180-kilometer-wide crater named Chicxulub (say it Mayan style: "CHIK-suh-loob"). It means "devil's flea." I guess they can be ornery. Some earlier extinctions may have also been the result of an asteroid or group of asteroids hitting the ocean, leaving little evidence for us to find.

One afternoon, I met Walter Alvarez, the scientist who developed and championed the theory that an asteroid triggered the K-Pg extinction. We had a delightful lunch. He is a thoughtful, enthusiastic guy who loves to teach. He also has a rare ability to look at the world and see things that other people don't. He's a bit like Darwin in that way. Walter and his father, geologist Luis Alvarez, proposed the asteroid impact idea in 1980, at a time when most of his colleagues thought that impacts were unimportant in the history of geology. The theory was considered highly controversial. Since then it has been carefully evaluated. It is now considered, by almost all accounts, very reasonable and likely.

When Earth was forming and was comprised of molten minerals and metals, the heavier materials sank to the center. Geologists would not have expected to find much iridium—a hefty element, atomic number 77—in the rocks at the surface. And generally they do not, but it does show up in one distinctive layer: in rocks that are 66 million years old, formed just at the time of the extinction. Alvarez reasoned that the iridium came from an asteroid, because asteroids are rich in iridium compared to Earth's crust. Apparently the asteroid hit near Chicxulub and disintegrated, spreading its debris all around the world.

Regardless of where they strike, large asteroids would boil seas, fill the air with dust and acidic compounds, and perhaps induce carbon dioxide to cook off out of the rocks and into the air, triggering a strong greenhouse effect, all of which in turn would change the world's climate faster than living things could adjust to. Giant impactors could create enormous waves in the ocean and in the atmosphere that could upset weather patterns around the world for extended periods. Perhaps asteroids have also helped unleash Earth's internal heat and caused subsequent volcanism. There is a joke in the space science community—not a joke, really—that the dinosaurs died because they didn't have a space program, so they had no way to save themselves from that asteroid.

Interestingly, many other large asteroid strikes did not seem to cause mass extinctions. Something about the Chicxulub impact was particularly bad. It was an unusually large impact, for one thing, and it may have struck in a geologically sensitive area. But those volcanoes in India are probably also part of the story. When the asteroid hit, the planet's ecosystems were already under stress. There is a great lesson for us here. Ecosystems can change only so fast, and the more insults you throw at them the harder it is for them to keep up. Human activity is causing another big ecological imbalance right now. How much of life on Earth will be able to keep up?

There is a great lesson to be learned from our neighboring planet Venus. Venus is very much like Earth in size and composition, but its surface temperature is about 460° C (860 F), hotter than your oven when it's set to "broil." The difference between the temperatures of Earth and Venus is not because Venus is slightly closer to the Sun. No, Venus is hot primarily because its atmosphere is full of carbon dioxide, a greenhouse gas that keeps the Sun's heat trapped in the planet's atmosphere. Venus is the extreme case of climate change: There is no way life, as we know it, could survive at those beyond-broiling temperatures. It would take a big change in Earth's geology and chemistry for it to become exactly like Venus. But humans are pouring carbon dioxide into Earth's atmosphere right now at an alarming rate, shoving our climate in that high-carbon direction, which is a terrifying prospect. We do not want to become even a little like Venus.

We know that a little cooling can cause a mass extinction as well; that's apparently what happened in the Ordovician-Silurian extinction. The key idea is simply that sharp climate swings evidently contributed to the Permian-Triassic and K-Pg mass extinctions, and probably to others along the way.

Industrial emissions are one way humans are changing this planet, but not the only way. We are also directly killing countless species at a rate that dwarfs the rates estimated in the previous Big Five extinctions. We are killing them mostly by destroying their habitats. We are forcing countless species to move, driving them from their ecological niches. The extra carbon in the air is holding in the Sun's heat; it's also soaking into the ocean, forming carbonic acid (like in soft drinks), which is compounding our looming troubles. The problem is not just that the ecosystems are changing; as many people note, conditions on Earth have been changing for as long as the planet has existed. The problem is the rate at which we are causing the changes. It's the speed that has us headed for the sixth mass extinction.

We can evaluate the current situation dispassionately, in evolutionary terms. If we destroy ecosystems, new organisms will take the place of those we kill. But as you might quickly realize, if we destroy the ecosystems we depend on, we will kill ourselves. Humans are smart and resilient. You might figure that a few of us will make it through no matter what happens. But how many of us won't? What will be the human cost, and the economic cost, along the way? How many of our genes, including your genes, will disappear forever?

The sooner we start acting to address climate change, the better. Don't take my word for it. Look at the eons, the eras, the periods, and the epochs. Will some future intelligent creature, digging through the ancient strata, try to figure out what happened to *Homo sapiens* back during the Great Holocene extinction? Earth will be here no matter what we do. Let's work together to save the world—for us . . .

15

ANCIENT DINOSAURS AND THE ASTEROID TEST

When I was in second grade, Mrs. McGonagle, our teacher, read from a large, authoritative-looking book and explained to us why the ancient dinosaurs went extinct. At the time, apparently, the best idea anyone had for their demise was mammals. Our ancestors somehow absconded with all of the dinosaur's food, or ate all of their eggs. Even as a very young person, I could tell that Mrs. McGonagle's heart wasn't in it. I could easily imagine an Ankylosaurus accidentally crushing a family of proto-rabbits with one stride while it was ambling around looking for a fruit snack. The intended message seemed to be that we humans were immune to extinction because we were on the winning team. But Mrs. McGonagle clearly could tell that the idea just wasn't reasonable, and so could I.

Today, we understand a great deal more about the demise of the ancient dinosaurs. We know they weren't evolutionary mistakes; unsuccessful adaptations get weeded out rapidly, but the dinosaurs stuck around for about 160 million years. (Human beings have barely made it a thousandth of that.) We know that mass extinctions are the result of environmental changes on a global scale. And we now have strong evidence that the ancient dinosaurs got walloped with a particularly extreme kind of change: not the hungry nibbling of some small mammals, but a long string of noxious volcanic eruptions topped off with a giant flaming rock falling from the sky.

Let's take a closer look at that asteroid that struck at the end of the Cretaceous. A 10-kilometer-wide rock might not seem all that bad, considering it was going up against our 13,000-kilometer-wide planet. But the asteroid was probably moving about 20 kilometers per second, or around 45,000 miles per hour. At such speeds, it carried the energy of a thousand billion (a trillion) tons of TNT. The scale of this is hard to imagine. This impactor must have tossed debris right up through the atmosphere 200,000 kilometers high, halfway to the Moon. Our planet was surrounded by a cloud of red-hot rock for days or weeks. Some of this material stayed aloft and blotted out the Sun. Some came crashing back down and set the world on fire. Sea creatures were cooked. And the ancient dinosaurs were either immolated where they stood, or they could not find food for their next meal. Meanwhile, our distant, distant ancestors were holed up in caves and burrows, and here you and I are.

We share a great deal with those ratlike ancestors. We have hair. We breathe air. Our females produce milk to nurture our offspring. We have four limbs, stereovision, and five appendages at the ends of each limb. What's not to love? . . . And all of this not only because a medium-large asteroid helped clear the way for you and me . . . but also because we haven't run into another asteroid that put us through the same infernal test.

As we look into the night sky with our stereoscopic vision system, we wonder whether or not we are alone in the cosmos. As I often remark, if you meet someone who insists that he or she has not wondered about our being alone, they're lyin'. Everyone has wondered about this fundamental question. So, how about this: Perhaps the reason we have never heard from someone or something from another world, from another civilization, is that those living things, somewhere out there in the cosmos, failed to pass the asteroid test.

You and I have the good fortune to live on a planet that is equipped with a large moon. In our case, we spell it M-o-o-n, capitalized because that is its name. We also happen to have had two superpowers that came into existence after a series of world-consuming conflicts. Through a terrible act of disenchantment, one world leader was killed, and his policy of racing to the Moon became a nation's policy, which drove members of our species to create space programs around the world. As a result, when an asteroid that could once again destroy the dominant form of life comes along, that dominant species (you and me) is ready to do something about that rock or block of ice.

We have the technology to prevent another mass extinction. We could give the asteroid a nudge, and life here would go on as usual. There are a number of relevant technologies under study right now. We could ram a rocket right into it, strap a rocket to it, or redirect it with a bomb. If we had enough fuel, we could pull it with the gravity of a massive spacecraft. At the Planetary Society we propose harnessing the energy of sunlight to zap it with lasers and nudge it onto a safe path. The ancient dinosaurs couldn't do any of these things—as far as we know. We can.

This is a train of science-fiction-style thought. But, I emphasize that it's not crazy. It's extraordinary, but hardly unreasonable. It is the study of evolution, and doing our best to learn about where we all came from takes us down this giant-impactor-from-space path. It's part of the process of science. It tells us something important about how we evolved, and it tells us something crucial about how to make sure we survive.

Asteroid impacts, by any reasonable reckoning, are the only preventable natural disasters. So my fellow Earthlings, let's get busy and see to it that we never take this kind of hit.

16

PUNCTUATED EQUILIBRIUM

If you've never been, I encourage you to visit and tour the sandstone slot canyons in Zion National Park, where the history of Earth is laid out before you, fine layer upon fine layer, like the pages of a book. If you want to sort out the story of how new species appear, and how other species go away, this is an excellent place to look. It's mesmerizing—just stand there and try to count the layers. The formations run from the Late Permian to Early Cretaceous Periods. That's 200 million years' worth of deep time. When you look closely, the layers are stacked like pieces of paper in a copier tray; only this stack is 1,000 meters—more than half a mile—high.

When I looked at these layers in 1997, while shooting the *Science Guy* show, I had the distinct feeling that Earth's history was a steady business. Each exquisitely fine layer seems to have been laid down in regular fashion. Wind carried the grains there, forming enormous dunes. From time to time, things got wet. From time to time, things got dry. When things were wet, the minerals calcite, $CaCO_3$ (carbonate with a calcium atom hooked on), and hematite, Fe_2O_3 (iron oxide, also known as rust), dissolved in the ancient water and cemented every sand grain into place. The rust makes for beautiful red tones. The scene mesmerized the crew and me with its steady drumbeat of sandstone building.

I am not the only one to form first impressions like this. Charles Lyell, the famous nineteenth-century geologist, had a similar impression. *Natura non-facit saltum*—nature does not make leaps—was the common wisdom at the time, and the belief in the steady unfolding of geologic history was called uniformitarianism. To his credit Lyell realized how much time he was dealing with. Basically, it's a lot more than you probably think. In fact, for most of us, it's more than we can imagine.

Uniformitarianism struck Darwin as being the way of nature, geologic or otherwise. He and his contemporaries figured species came into being at a pretty slow but steady clip, and had been doing so since the beginning of time. But is it true? That turns out to be a very important question if you want to know where new species come from—if you want to find a way to combine the ancient fossil evidence with the modern genetic evidence and take Darwin's ideas into the twenty-first century.

As Darwin and countless subsequent researchers examined all the fossils they could find, they ran into a persistent puzzle. There seemed to be a lot of missing animals and plants. Many of the key fossils that you'd expect—early types of birds, for example, the transitional forms—just weren't around. Darwin called the missing fossils, ". . . the most obvious and gravest objection which can be urged against my theory . . ." It was a puzzle a lot of scientists wanted to solve, and remained so for a long time. The subtlety of the eventual answer throws millions of people off even today. Creationists and uninformed people everywhere still harbor doubts about evolution partly because of the concerns about missing fossils that Darwin and other nineteenth-century investigators had early on.

Some of the problem was a lack of information. In Darwin's day, there were far fewer fossils available to study than we have today. The museums and the vast collections we have now didn't yet exist. There was no economical way to share the specimens that did exist, no transmission of electronic images between handheld screens, and so on. Most of the fossils scientists expected to find based on their studies had not yet been unearthed. This is especially true of fossil remains that were to connect us humans to an ancestor common with great apes, bonobos, and chimpanzees. This hypothetical single fossil or set of fossils came to be called the missing link or links. I remember well as a kid the derisive description of someone that you thought was uncouth, illiterate, or culturally inept as being a "missing link." My parents referred to one of my older sister's boyfriends as the missing link. They felt he was an unsuitable suitor. I am pretty sure he was a human like the rest of us, and not a missing link in the fossil record. My old boss on the other hand . . .

During the decades after Darwin published, field geologists and paleontologists have gone on to uncover and discover thousands and thousands of fossils. They have uncovered an astonishing number of ancient dinosaurs, an overwhelming assortment of long-gone mammals, and an uncountable number of sea creature fossils. Just two years after Darwin expressed his concern about the missing fossils, the famous fossil of the birdlike *Archaeopteryx* was found in Germany, and that's just one example. Later, fossil hunters found a whole range of human ancestors, including *Sahelanthropus tchadensis*, which may in fact be a shared ancestor with chimps as well. You could say that every one of these is a no-longer-missing "missing link."

The issue of missing links has been kept alive mostly by people who believed (and even now believe) that Earth is no more than ten thousand years old and that humans are unique, with no ancestral connection to any of the billions of living things that came before us. By insisting, quite falsely, that there was no known transitional form between apes and humans, they introduced doubt into an enormous number of minds.

Even after the so-called missing links were found, though, some of the big mysteries that troubled Darwin lingered on. If anything, filling in the fossil record made them even more troubling. First, new species seem to show up pretty fast in the geologic record. Darwin pondered this problem when he wrote: ". . . Why then is not every geological formation and every stratum full of such intermediate links . . . ?" Second, once a species is established, it and its descendants often hang around, or hang upward into the rock strata, for a long time. The trilobites alone survived in various incarnations for more than 250 million years. Somehow, evolutionary change seemed to happen very fast, but also very slowly.

Making sense of that paradox required blending the latest knowledge about modern ecosystems with the study of creatures that lived long, long ago. It wasn't easy to get everyone talking. As my colleague and friend Don Prothero wrote, "Meanwhile, systematists (biologists, who study the naming and relationships of organisms) were busy describing new species, but few thought of the evolutionary implications of their work. There was simply no common thread among them, and there appeared to be no way to show that Darwinian natural selection was compatible with genetics, paleontology, and systematics."

This challenge was tackled brilliantly in 1972 by two young (but now very well-known) evolutionary biologists: Niles Eldredge and Stephen Jay Gould. They did compelling analysis of a tremendous number of fossils and came to realize that, although we have a great many fossils that show us big lines of descent, there is a surprising absence of fossils that would tie certain of these lineages to other lineages. It still wasn't obvious exactly how dinosaurs became what we think of as modern birds, even once the overall course of that evolution was quite clear. Similarly, it wasn't obvious how fish ended up walking on land, or how land animals went the other way and ended up swimming around as air breathing fluke-thwapping whales and smiling dolphins. Some of life's biggest transitions seem to have happened so rapidly that they disappeared between the grooves (or digital bits) of the fossil record. That's what Eldredge and Gould set out to explain with a spectacular new extension of Darwin's ideas.

You may have heard the phrase they coined for this phenomenon: "punctuated equilibrium." I met Stephen Jay Gould at a small group dinner, and I can bear witness that he had a helluva vocabulary. Along with his command of English, he seemed to be nearly fluent in Latin. At any rate, punctuated equilibrium ("punk eeck" among slang-talking paleontologists) has caught on as a description of the mechanism that produces species. Someone like me might have called it "cutoff change," or "isolation speciation," or "genetic island formation." For me, it might be: "I's all good, less the creek don't rise." Which might be expressed *Nybonically* with better grammar: "If the creek rises, a population may get isolated on a genetic island" (whimsical *Nybonic* expressions have or pertain to the characteristics of the way in which Bill Nye makes up words and phrases).

Whatever name you use, the important thing to know is that the answer to Darwin's puzzles lies in the size of populations—specifically in small, isolated populations. Darwin had pictured one whole species giving way to another. That's how it went for Darwin's famous finches on each of the Galapagos Islands, for example. Once you let go of that old uniformitarian way of looking at things, the situation becomes a whole lot clearer. When a small group of organisms gets isolated (in an isolated patch of forest, on the other side of a rising creek, etc.), some individuals are more prone to form new species. In a small group, any mutation is a much bigger part of the mix, and a successful mutation is immediately a much bigger deal.

Since the landmark 1972 paper by Gould and Eldredge, many studies have been done both with real populations and with mathematical simulations. The results explain why evolution appears both fast and slow: It *is* both fast and slow. Large populations tend to stay genetically about the same. Paleontologists say populations tend toward stasis. Small ones can diverge quickly into new species. Now that you've read this explanation, I hope your response is something like: "Well, obviously . . ." This could also be expressed here in the early twenty-first century with a single syllable dripping with sarcasm: "'Cha . . ." It seems to have derived from: "Well, yeah . . ." Keep in mind though, it was not obvious to many people in the preceding one hundred years or so.

The significance is profound in the context of all that came before in evolutionary thinking, and with the distraction of creationists trying to teach science students kooky ideas about the natural history of Earth even today. Punctuated equilibrium explains why we are missing a great many transitional forms in the collections of fossils kept in institutions around the world. Let's say we come across a string of islands (the Galapagos) that form an archipelago in the eastern Pacific Ocean off the coast of what is now Ecuador. If the weather conditions are violent enough, animals from the mainland can find themselves blown or washed onto these islands. (I talk more about that elsewhere in the book.) These islands are close enough to the mainland to allow organisms to get blown or washed there, but they're too far away for the isolated groups to have a significant number of encounters with the old tribe or flock once they've landed. Their populations are isolated. Evolutionary biologists often use the term allopatric, from the Greek for "other fatherland," or "other homeland."

Compared with the birds back on the mainland, the finches on these islands are members of a pretty small tribe, or flock. If one of them happens to be hatched with a beak that's just a little better than his neighbor's beak for cracking nuts, the bird with the superior beak has a better chance of getting square meals. The significant thing here is that his better-beak genes are a larger fraction of the flock. His better-beaked babies' genes will become a bigger bucket in that island's gene pool. As I say, it seems obvious once you know the answer. But where we really see it is in mathematical models. Here (or there) you can speed up evolution with electronic computer simulations. The effect of punctuated equilibrium jumps right out. It's the reason we just don't see many of the transitional fossils. There are inherently fewer of them, and the changes happen quickly. Very few of those intermediate organisms get preserved for us to find millions of years later.

If a small population gets just a slight advantage, that small population can become big. Since the population we're talking about is isolated, it can become just different enough from its ancestral tribe or flock or school for its individuals not to be able to successfully mate with the old gang. Those individuals are of a new separate species. When we look at the fossils of the new separate big bunch, we don't see the intermediate linking individuals, because there were so few of them. Once you understand genetic island formation or punctuated equilibrium, it would be weird if things were any other way. The missing nature of missing links is actually further proof of evolution. It's just what we expect to find out there in nature. If the fossil record were perfect—now *that* would be a mystery.

By the way, while I'm writing here about the incompleteness of the fossil record, keep in mind that the incompleteness is becoming less and less incomplete. Every week or so, paleontologists find another amazing animal whose remains were locked in rock. There was a recent discovery of a two-and-a-half-meter-long (8 foot) millipede fossil from the Carboniferous Period, about 300 million years ago. The fossil is well-enough preserved that investigators can see that it was a vegetarian by examining its very long intestinal tract, which is now preserved as solid rock. I've been to Ashfall State Park in Nebraska and seen the seeds preserved in the tummies of two-ton, long-extinct, North American rhinos. We always seek more fossils to learn more about the past, but we sure do have a great deal of information at our disposal these days.

So far this discussion has focused on change (after all, that's where the cool stuff happens), but stasis is a major characteristic of evolution as well. Populations in ecosystems tend to stay in balance. Why wouldn't they? If they stay in the same locale, and have the same amount of sunlight and food resources for years on end, individuals are born and die, while in the bigger picture, things remain about the same. Once in a while, you'll hear people refer to an organism as a "living fossil." Even I have used the term. While I'm sympathetic to the intent, it's a nonsensical expression. Fossil refers to something that has been dug up. If it's alive, it's not dead. . . . That aside, I think I know what someone means, when he or she refers to a living fossil. They mean an organism that has been unchanged for a long, long geologic or evolutionary time.

You may have seen or even own a nautilus shell. These are the wonderful shelled sea creatures that move from one chamber to another, building as they grow, in a logarithmic spiral. They have what we might describe as pinhole cameras for eyes. And, they've had them for the last 500 million years. The animals alive today are not fossils. But, they are just like their ancestors, who are. You may have seen pictures of the coelacanth fish. It was thought to be extinct. Only its fossil remains had been found until 1939, when a population was discovered off the southern coast of Africa living as their ancestors had lived for the last 65 million years. By the way, both the nautilus and the coelacanth are endangered species—thanks to humans. We're killing them off for their shells and out of curiosity. We may indeed soon make these living animals into dead fossils, a troubling and permanent condition.

For animals like nautiluses and coelacanths to live unchanged from generation to generation, they have to live in environments that don't change too much over long spans of time. They keep accumulating mutations in their genes, yes, but the environment's stability does not give major changes an advantage. It's not a coincidence that these animals live in the ocean. There, you have a much better chance of swimming around in an unchanging environment; leastways you did, before humans showed up. From time to time, you have probably used sentences that include the phrase "balance of nature." But when there's a big change—say a volcano erupts in your neighborhood, or an enormous storm blows you out to sea and you land on a happily unpopulated island—populations get isolated. That's when things can happen fast.

I lived in the Pacific Northwest for many years. I still visit often. The smell alone is enchanting, not to mention the alpine vistas and fecund bays and inlets. There was a great deal of controversy concerning a particular bird that, to many, had no great economic value and was not of any particular interest to people who wanted to cut down ancient trees for lumber. The ancestors of these birds lived in the conifer forests, flying around and eating voles and mice. Then humans showed up and started cutting down trees like crazy. Humans wanted the wood to build houses and structures for commerce. The old-growth lumber is the best you can get. The lumber is fine-grained with little wane. It's beautiful to look at, and it's ideal for the construction of strong buildings that are stiff enough to handle high winds, yet compliant enough to flex their way through a powerful earthquake now and then.

Well, if you're one of the spotted owls, *Strix occidentalis caurina*, this is bad news . . . Humans are coming to cut down your home. If you can manage it, have your offspring, come up with a clear-cut way to make a living in a clear-cut, or you'll go extinct, which is what the

Northwest spotted owl may soon be. We are changing environments around the world. We are changing Earth's climate. We are causing an extinction of an astonishing number of species. If history serves as a guide, new species will arise to take their places, but the rising will be on geological timescales. We're the only animal out there that can turn this rapid change around.

Populations of living things tend to remain in equilibrium, but now and then the whole ecosystem runs into an exclamation mark; certain individuals and populations hit a full stop. That can change things in a hurry. The story of evolution is one of equilibrium punctuated with big changes. We have a strong interest in minimizing the full stops and emphasizing life's run-on sentences, for the sake of all the living things that are not humans yet on whom we depend for a healthy world—our world, and the world of our progeny.

17

CONTINGENCY, BOTTLENECKING, AND FOUNDING

How do different species originate? It's the original question. When Charles Darwin was piecing together his ideas, nobody had any knowledge of DNA or genes. He was able to deduce the principles of natural selection, but he was almost completely limited to life's external and internal shapes. Today, we can look into the code of life inside every living thing. Now we can get at the mechanism of evolution. We can look at the molecular record that documents the inside story of how new species emerge and drift away from each other over time. A revolution in gene mapping has sparked a revolution in evolutionary theory as well.

In 1973 the Ukrainian-American geneticist Theodosius Dobzhansky wrote a compelling essay, "Nothing in Biology Makes Sense Except in the Light of Evolution." He is generally credited with starting the discussion or intellectual dialog often called the "new synthesis" of evolution. Dobzhansky incorporated the biochemical details and role of the technical description of a gene: the specific sequence of nucleotides (aka, the genetic code) that comprise a portion of a chromosome. Described this way, a gene is a construction plan that ultimately determines the order of amino acids needed to create a specific protein. Simple enough? Actually it is fantastically complex, and biologists are still learning the details of how it works. However, this molecular point of view is absolutely, completely, in every way consistent with the observations and conclusions that Darwin made: DNA directs the construction of strings of chemicals; those chemicals influence the configuration of the whole organism; that configuration influences how likely it is that the organism will reproduce and keep spreading more copies of the code.

Dobzhansky's influence was so profound that we often forget about it. He tied the phenomenon of gene mutation—the natural mistakes that happen in making copies of elaborate or complex molecules—to the happy accidents that become, in Darwin's term, the favored descendants. If the mutation was of value to the offspring in reproducing, that mutation will get passed on and on.

Before this synthesis, the means by which a species became separate or a new species emerged was not exactly clear. Investigators had presumed, quite reasonably, that every individual of a given species had pretty much the same genes. What made one different from another was thought to be connected to genes, even though the variation among individuals within a species is not very large. The reason you and your sister have different colored hair was understood to be genetic but a small fraction of one's inheritance. With the twentieth-century "modern synthesis," it became clear that every feature of an individual is expressed in her, his, or its genes. We take this idea for granted now. It gave scientists an understanding of what it takes to become a species. The genes mutate enough through enough generations, and you get individuals that can no longer reproduce with each other; they've become separate or a new species.

The most important insight from the modern synthesis is that it enables scientists to understand how populations split into different species. It identifies one of the key mechanisms behind life's diversity. Let's dig deeper by thinking about a population of beetles. Believe me, there is no shortage of different beetle populations; there are 350,000 known beetle species, and probably plenty more left to be discovered. Let's say they live in a forest. One year in the mountains above our forest, there is a significantly higher snowfall than there has been in decades. As the snow melts, a larger than normal flow of water cascades down a major hillside forming a new river right through the beetle's natural valley habitat. The beetles, who happen to make their living in the forest litter or duff below the trees on the forest floor, are now separated into two communities, the left-bank beetles and the right-bank beetles. They no longer communicate and conduct beetle transactions, including egg fertilizing and egg laying.

With imperfect copying of the genes in the two populations, generation after generation, eventually the lefters and the righters have genes that are too different, and the two groups can no longer reproduce with each other. The two populations, in our simplified but not unreasonable example, became distinct species. Meanwhile, other natural selective forces are at work as well. The lefters may live on a portion of the river where this same out-of-the-ordinary snow event flooded the banks and killed off a large swath of trees by drowning, essentially suffocating, their roots. Certain individuals in the left-bank population may have chanced upon a slightly different jaw or mandible that makes them ever so slightly better at chewing the dead wood of the suffocated trees. They will be better nourished. Their eggs may hatch under better conditions, such as freshly chewed-out galleries that keep the eggs at a more productive temperature than their fellow tribesman (tribeetles?). So, their eggs do a little better, and more left-bank beetles are born with better-suited mandibles and gallery-chewing capability. Meanwhile, the right-bankers carry on as they did before the flood. Their environment hardly changed at all, and yet the left and right populations diverge.

There are two more consequences to this sort of turn of events having to do with populations of organisms becoming isolated or cut off. In our example of the beetles and the catastrophic river-forming flood, it is quite likely that the two populations will not be of the same size. In my imagination, I see the left-bank beetle community being much smaller. Their territory was overrun by the melting snow flood. The right-bankers, literally on the other hand, were on the inside lane of the river bend and very few of them were drowned in the great snowmelt, river-forming flood. They have almost as many individuals as before. When it comes to insects, the numbers can get huge. If you've ever been in Minnesota in July, the flies . . . well, they are uncountably numerous and ornery. If there were only 10 million, a Minnesotan canoeist's life would be a lot less troublesome. With that in mind, let's say we have 10 million beetles on the right bank all carrying essentially the same genes as they had before the flood.

On the left bank, we have only ten thousand beetles who made it through our imagined cataclysm. Beetles being the way they are, they reproduce like crazy, and the ones with the better-suited-to-chewing-dead-tree jaw do especially well. On top of that, their predators, which may have included other species of insects like praying mantises and birds such as wrens, would not have responded to the novel landscape. They may not be as good at finding beetles in the muddy zone full of dead tree trunks, which would enable the new style of beetle on the left bank to reproduce with even more success. For this thought experiment, let's say the populations eventually stabilize. The predator and prey relationships settle out to stable numbers of each. Even if the left-bank population reaches the same size as the right-bank population, their genes on the left bank will have a great deal less variety, because they are all descendants of a much, much smaller population. This is called a genetic bottleneck, because all of the beetle ancestors passed through a narrowing of their genes that is analogous to the narrow neck of a bottle.

Here's the wonderful thing about bottlenecks. We can measure the genes in different populations and infer a great deal about who begat whom in the plant and animal kingdoms. From those inferences, we can draw conclusions regarding the natural history of whole continents and ecosystems. With the tools of modern gene sequencing (including some amazing chemical reagents and extraordinary machines), we can sample the DNA from the blood of the different beetles and determine which population has more genetic diversity. The population that survived the flood would have less diversity than the population that was largely unaffected. This is what we would expect. But it's important to keep in mind that it was possible to understand the outlines of this process long before anyone knew about DNA and sequencing.

A powerful, real-world example of a genetic bottleneck took place in the Galapagos Islands. A young rambling Charles Darwin noticed, among other things, that the island birds, specifically the finches, were all very similar. Nevertheless, they also had slight but distinctive differences in their beaks. Darwin realized the implications of what he was seeing.

Let's say you are a happy finch flying around on the mainland of what is now Ecuador, and an enormous storm blows through, a cyclone a few hundred kilometers across. You and several members of your community, your flock, get caught up in the high winds, while you are idly tweeting about varieties of nuts. You all get blown out to sea. Many of you become too tired to fly and disappear beneath the tumultuous waves. But a few of your comrades and you end up on an island. There are plenty of nuts, because apparently some nut tree seeds had been blown here during similar storms many years earlier. The nuts are good enough to eat. You and certain of your community members have beaks that other birds often chided you about. You and a few others have a bit of a hook on the ends of your beaks. These hooks work great for knocking nuts open. You form a new community and reproduce for many years. Your descendants several generations hence fly about and alight on rocks and branches discussing nuts and what the neighbor finches are up to.

Then another enormous storm blows through this island. Major storms are quite common in that part of the world. Despite the experience of the ancestors, or perhaps even because of it, these great, great, many times great, grand-birds also get blown out to sea to the west, this time many drown, but a few alight on a new next-to-the-west island. A new population gets a new start, and so on. With each successive catastrophic storm event and landing on each successive island in the Galapagos Archipelago, we have a population or community that came to grow and multiply from a set of genes that inherently has less and less diversity.

Darwin came on the scene long before anyone knew about genetic diversity. A great many researchers since his time have incorporated the new ideas into his theory. The gene sets of the finches are what you'd expect if this were what went down (or flew up) in this part of the Pacific Ocean. Along that line, there are iguanas in the Galapagos region that are reminiscent of those on land, but quite different. They can swim, for one thing. Jungle iguanas have no such skill. These lizards didn't get to the islands by flying, but storms are occasionally powerful enough to get a few of them out there by other means. Certainly a great many logs or fallen trees end up in the ocean along the heavily forested coast. If you're a lizard clinging to a tree, what are you going to do if it ends up in the sea? You hang on. If you end up on an island, you do what you can to make a living and participate in—we presume hot, albeit deliberate—lizard sex with another drifting tree-mate that made it out there by the same means. We can measure their genes and compare them with the genes of the lizards back on land and get just the result that evolutionary theory predicts: Less diversity from east to west along the archipelago.

We call these sorts of novel populations that take hold and make a living in areas that are new to them "founders." They found a new community, just as human founders start companies or institutions. Only these animal founders fought for their lives and especially the lives of their descendants.

In general, founders arise from, and give rise to, genetic bottlenecks. Because there are fewer individuals that make it to a new area to found a new community, there will just be less diversity in the genes of that smaller founding population than we would find in the native tribe or community whence they came. An oft-cited example of this is the Afrikaners in South Africa. They came from Holland, and apparently carried a gene that makes a human susceptible to Huntington's Disease, a degenerative brain disorder. People who suffer from it exhibit jerky movements especially in their faces and shoulders, and ultimately suffer from a type of dementia. There are a great many more Huntington's patients in the Cape of Good Hope region of South Africa than in the rest of the world's population. The cities of Cape Town and Johannesburg were founded by a relatively few Hollanders, some of whom carried this troubling gene. The genes of the Afrikaner population in southern South Africa passed through a genetic bottleneck, when their communities were founded. Another famous example is the prevalence of hemophilia in the inbred British royal family.

The phenomena of bottlenecks and founders have led scientists to speculate on the role of contingency or happenstance in evolution. Scientists have long debated the importance of new environments in biological and especially genetic diversity. I like to put it this way: Do we need catastrophes big and small, including the devastating mass extinctions, to make new species? The beetles in our earlier thought experiment would be an example. It's a fascinating question, because it takes us back to the deeper question: Where did we all come from? Put yet another way: Would we all be here if all hell hadn't broken loose here on Earth a few times?

Now that geologists know what they're looking for and have designed the instruments to go looking, they have discovered dozens of meteoric impact craters around Earth. The dinosaur-menacing rock that struck Mexico 66 million years ago was hardly the only one. You have to figure that if you're a living thing, like our beetles in a forest, and a giant white-hot rock lands nearby and sets everything on fire, there's going to be trouble. You, and every beetle in your community, may get burned to Coleopteran crisps. Or perhaps, you have the good fortune to be on the edge of the inferno. You and a few comrades live through it, and before you in the aftermath is a whole new landscape to explore with very few living things upon it. Furthermore, you may have very few competitors for some time, perhaps decades. Even if you don't live to explore and eat the shoots of the new plants that grow in the newly exposed nutrient-rich soil, perhaps the eggs you managed to lay that first season do survive, and your offspring have the run of the new landscape for years. A huge successful population gets established, one with much less diversity, but still carrying your genes, and so on and on and on.

And as we've seen, asteroid impacts are not the only catastrophic challenges life has faced. Enormous flood-basalt volcanic eruptions, similar to the ones in Siberia and India, have poisoned the globe with dust and noxious gases at least fifteen times during known geologic history. Global cold spells may have led to periods when the entire planet was encased in a casket of ice. The motions of the continents have repeatedly triggered drastic changes in climate, ocean circulation, and ocean chemistry. There were probably other grave insults we don't even know about yet. And there were certainly countless smaller but still influential environmental crises (droughts, floods, etc.) that wiped out populations and created the right conditions for new founders to emerge.

The question for evolutionary biologists such as Dobzhansky and, more recently, the influential American researcher Stephen Jay Gould, was or is how many such events do we need to account for the diversity we see in the world today? How much contingency do living things need to get mutations sufficient to account for all of us Earthlings?

Some evolutionary scientists argue that we'd all have four limbs and a jaw whether there were catastrophes or not; that would be convergent evolution. They claim that life on Earth would look roughly the way it does today no matter how many happy accidents occurred, or didn't occur. Some argue that a planet just has to have catastrophes in order to provide new places, new so-called ecological niches, to create diversity. They point to the mass extinction events in the fossil record and argue that without those worldwide resets, we wouldn't see the animals, plants, and microbes we see today. You can take it down from worldwide to local and argue the same thing. Without big changes in the environment or the predator-prey relationships, we would not have diversity.

Why this debate gets some very, very well established biologists worked up into such a tizzy is a little unclear. It's clear that we need both convergence and contingency. We need to obey the laws of physics and we need to take advantage of opportunities. A look at the species living Down Under shows you what I mean. In Australia, we find counterparts to dogs and cats in the natural ecosystems. There are scavengers that travel in packs, and there are nocturnal hunters that seek food while stalking solo.

Would we naturally have an Australian counterpart to a dog and to a cat if the Australian continent hadn't been connected to the mainland by a land bridge for some time, during an ice age, let's say, when ice and snow held water up and out of the sea? Are a pack hunter and a night stalker needed in every ecosystem? Are they a natural consequence of multicellular animals feeding off plants and things that feed off of plants? Are those ecological niches inherently filled by the nature of life on Earth, or are they just the chance descendants of descendants? If they had been cut off earlier in geologic history, would there be no such counterparts?

To me, it's obvious that both effects are in effect. We cannot say whether four legs or eight or dozens are inherently the best solution to mobility problems. Instead, we can say that all of the schemes we find extant today are a result of solving the same physics and chemistry problems by playing the hand each of us Earthlings was dealt. But I see no evidence that there was a superpower (a super-dealer?) running the show here. Come to think of it, there is overwhelming evidence to suggest that there isn't, at least when it comes to picking winning and losing body plans and designs.

You can look into this topic further and decide for yourself which factor is more significant: Is it convergent forms like wings and feet, flowers and stems? Or is it contingencies like floods, asteroids, and shifting ice caps? I don't see any reasonable hypothesis that explains things as we see them that doesn't include both convergence and contingency in nature. How else could it be? If things were any other way, things would be different.

18

MOSQUITOS IN THE TUBE

The problem with a lot of the evidence of evolution is that it is hard to *feel.* No matter how crisp and clear the ideas may be, even the most beautiful fossils remain static chunks of petrified bone, and DNA is still a mess of invisible molecular code. Wouldn't it be something to watch a new species emerge in front of our eyes, over a single human lifetime—to experience evolution not just as an intellectual analysis, but as an immediate phenomenon that could fly right up and bite you? Well, yeah. That's why I got excited when I discovered a place where that is really happening. The location of this amazing evolutionary showcase? The London Underground, affectionately known as the Tube.

In London, as in so many places we visit, there are mosquitos. In biology, we say class: Insecta, order: Diptera, family: Culicidae (insects with two wings, who are like gnats), genus and species: *Culex pipiens* (they buzz). Unlike many other places in the developed world, London used its subway stations as bomb shelters, back during World War II when the city was being attacked with rockets from Germany.

To make them harder to intercept, and to make the whole Nazi campaign that much more demoralizing, the rockets were often launched at night. Londoners avoided collapsing buildings and flying bricks by going underground to the Underground. Thousands of people slept in the tube stations as the bombs fell above. The citizens were attacked from the air—not only with exploding V-1 buzz bombs, but also by buzz-biting mosquitos that followed them into the subway stations. There was nowhere for the citizens down there to run, so the mosquitos had a field day . . . or field night. Mosquitos generally feed at night, especially at dusk, and feed they did.

The layout and construction of tube stations is such that there are often puddles of water between the tracks and near the cracks. Mosquitos lay their eggs in water . . . perhaps evidence of their waterborne ancestry. With plenty of blood to suck and a place to raise a family (okay they lay eggs and fly away), the London mosquitos had no real need or pressure to ascend to the surface world and look for birds or some other people to poke with their proboscises. A new population of mosquitos got established, one that lived underground, completely separate from their cousins up above.

An isolated population is where new species are likely to emerge, and that's exactly what happened. In approximately fifty years, a new species of mosquito has appeared in the London Tube: *Culex pipien* has become *Culex molestus*. Translating from the Latin roughly: the "mosquito that buzzes" has, by being an isolated population for long enough, become a new species, the "mosquito that bothers."

Experiments have been done in which the eggs of one of these two mosquito types was mixed with sperm of the other. Generally, they cannot breed together. It works just once in a while. We are watching, in just one human lifetime, the emergence of a new species. The annoying underground-only mosquitos came to be separate just by not being in the aboveground mix. As they reproduce—that is, as they mix genes and have more annoying mosquito offspring—their genes are not copied perfectly, so eventually, apparently, the undergrounders' genes are different enough to make them un-mateable with their very, very recent ancestors who still live above. Upstairs, the *C. pipiens* feed on birds and people. Downstairs, the *C. molestus* feed on us only. Still, you have to figure the molesters can find the occasional rat down there in the tube tunnels. For now, I guess we let them go at each other.

Although the London Underground mosquitos are becoming a separate species overnight, or almost overnight, as of this writing they are not quite a separate species, not yet. In laboratory tests, occasionally certain undergrounders can successfully mate with certain abovegrounders, even though a majority of each cannot successfully interbreed. Looking at the world with deep time in mind, you can imagine that in just another few decades they will be fully separate. They will be unable to mate with each other under any condition; their genes will be too different. Evolutionary biologists see an important lesson here. Because of mutations, populations diverge genetically as time goes on. We can infer that if we went back far enough, we would find a common ancestor for every living thing on Earth. That is one of Darwin's great insights. That is also the small message of the London mosquitos, writ large.

As you look at these two populations of mosquitos, you can see the divergence happening. You have to figure that the same process has always been happening, at least since life began here on Earth. Separate species arise from ancestors, from parents, from organisms that came just before, and just before those that were just before, and so on, and so on, and so on back into the millennia.

The two types of mosquitos still share a great deal of heritage—not surprising, considering how little time they've had to go their separate ways. But you know what? Humans are still genetically very,

very close to their cousins, the chimpanzees and great apes. We are different enough to be considered a separate species; humans do not mate with apes (at least, nobody I know). But anthropologists have found dozens of fossils of individuals who are neither like a chimpanzee nor like us. They're clearly related, with hands and hips and skulls almost like ours but not quite. Were they all completely separate species? Could any of them have bred with modern humans if they lived at the same time? These are not idle questions. They are the focus of some of the most fascinating evolutionary research going on right now.

You've probably heard a good deal about Neanderthal [Nee-AND-er-tall] people or almost-people. Sometime around 500,000 years ago, the ancestors of Neanderthals and modern humans diverged. It's a difficult business making inferences about ancient people from the limited remains available, often just a few fragments of bones and teeth. When I was in school, it was believed that the Neanderthal people were completely different from our kind of people. This is to say, it was believed that they and we were entirely different species and never mingled with each other, much less . . . um . . . got intimate. Now the evidence shows that our ancestors and Neanderthals not only were alive at the same time, but apparently they interacted. They may have traded with each other, but that's not all they did. Groundbreaking genetic work by Swedish biologist Svante Pääbo and others shows that the two kinds of protohumans had sex with each other. We were hardly more different than those two breeds of London mosquitos.

Somehow, we all got through it. Humans and Neanderthals were carrying on together as recently as thirty thousand years ago. Apparently, we have enough genes in common to have pulled this off. We were close enough to have had babies. I cannot help but hearken to the time-honored lyrics of the accomplished jazz artist Joanie Sommers, who recorded an uncharacteristic 1962 pop-chart hit with "Johnny Get Angry," in which she asserts that: "I want a brave man / I want a cave man . . ." Because we've found extraordinary artifacts in caves in Europe, we often associate ancient people with caves. Just looking at an artist's rendering of the Neanderthal physique, you might easily presume that these males were big and strong, which the protagonist in the song would find appealing. Undoubtedly they were, but Neanderthals also had and needed bigger brains than modern humans to operate their bigger frames. You and I have smaller brains than Neanderthals, but we have smaller bodies as well. Our brain-to-body ratio is just a little bit bigger than theirs was.

The lesson to be learned from our cave-guy and cave-gal ancestors, along with our mosquito nemeses in the London Underground, is that the expression "a separate species" is meaningful only in cases where the organisms diverged from a common ancestor some time ago. The process of evolution produces a genetic spectrum, and we're observing it among mosquitos right now in a big city full of modern humans. It reminds me of the big box of crayons that some of my classmates had in elementary school. There was "red-orange" and "orange-red." If you visit the Crayola Crayon factory in Easton, Pennsylvania, you can watch the pigments get mixed. You can see that it's just a little bit extra of red that produces red-orange, and a little bit extra of orange that produces today's mango-tango. The difference is slight but readily observable. In the case of species, the difference starts out slight, with just a few genetic mutations. But in the very long run of deep time, the difference becomes extraordinary.

The remarkable thing about the evolutionary process that produces new species is that it's brought on by small, random changes in genes. But then the resulting organism or offspring is confronted with an environment that makes or breaks it. The mutation might be random, but the selection pressures are deadly specific. With remarkable speed, new species are created, and diversity abounds.

Since the emergence of a new mosquito protospecies happened so quickly, I cannot help but wonder if there is a little something extra that the underground mosquitos chanced upon. Are they a little better at finding humans than they are at finding birds to suck blood from? Or was it just easier down there, never having to go looking very far for a host? Every day, the hosts come down the stairs to them. Did they just drift apart genetically? Or is there some other advantage they have? The answer may be a combination of these factors. Considering all that's involved, the rate of change is astonishingly fast. Just think what could happen if every living species could beget a new species every one hundred years!

Given that nature produces new species so readily, it's no wonder that we now have at least 16 million species on Earth, with many biologists speculating the true number is far greater. There might be 100 million species extant right now, with perhaps one thousand times that many having come into existence and disappeared. Even as they went extinct, they gave rise to the organisms alive in the less remote yet distant past, eventually giving rise to us and all the living things we see and dig up.

As you might expect, biologists can take the next logical step. They can infer how long ago what we now consider a species came to be separate, even if they don't have fossils to study. One of the most powerful ways to reconstruct ancestry is to assay the genetic code of two living species, observe and measure their current rate of mutations, and then run the genetic clock backward. The amount of genetic difference reveals how long the two species have been going their separate ways. In some cases, the geneticist can then compare notes with the paleontologists to see what the fossils show. There's some guesswork involved, so you wouldn't expect a perfect match, but most of the time the inferred ages come out very close.

That's another remarkable piece of evidence about how evolution works. Two totally different techniques, two totally different ways of looking at life, give the same chronology of the history of life. And it's abundantly clear with each observation farther into the past that all of us Earthlings have DNA, and we all came to be from a common ancestor—just as Darwin deduced.

In the case of the mosquitos, biologists measure the genetic differences between *Culex pipiens* and *Culex molestus.* They measure their rate of change, and sure enough the genetic clock matches up with a split that occurred right around the time of World War II. In the case of humans, we can measure the DNA of someone alive today, and in certain cases, measure the DNA of someone who died centuries ago. A company called 23andMe will analyze your DNA and reconstruct the broad details of your own personal ancestry. Taking this same basic approach, geneticists have identified widespread human genes derived from Genghis Khan, and tracked cancer genes that originated in the Middle East thousands of years ago and then spread with Jewish populations as they moved around the world. My recent ancestors are from Europe, but my DNA assay shows that I'm substantially Bantu.

By learning how our ancestors' DNA was configured we can get also some sense of where human evolution is headed now. As we'll see, there are other ways to figure out how humans are currently evolving. Our descendants may puzzle over how we missed some obvious aspects of evolution, ones that would have helped us make better decisions as a society.

Perhaps one of the most important insights is that humans are extremely uniform genetically. We are just emerging from our own genetic bottleneck. As a result, we are as much like Neanderthals as those two breeds of mosquitos are like each other. Everyone living today is much, much more closely related than that, even. Perhaps that sense of shared heritage can help motivate us to work together more often and accomplish great things.

19

CONVERGENCE, ANALOGY, AND HOMOLOGY

People love to find patterns. It seems to be hardwired into our brains. We do it all the time. We say, "this flower looks like that kind of flower," or "that cloud looks just like a dragon." Perhaps the ultimate in groupings is the time-honored phrase, "tastes just like chicken." Pattern recognition undoubtedly aided our survival by helping us recognize good foods, dangerous predators, family members, and so on. When it comes to understanding evolution, though, this tendency has some strange consequences. It tricks us into seeing relationships where there are none, and missing ones that are not written obviously on the surface. On the other hand, it intuitively exposes the physical laws that shape so much of natural selection. I suspect that you know things about evolution that you don't even know you know.

The human penchant for pattern seeking is what led early naturalists to group organisms by how they appear. It was a start in making sense of life's bewildering diversity. You don't have to be an especially disciplined naturalist to infer that although they both fly, bees are fundamentally different from birds. You could also probably easily come up with the idea that birds and bats are somehow more closely related to each other than they are to bees. With just one more step—or flap—you can tell that penguins are more closely related to crows than they are to crawfish, even though two out of those three swim. This impulse toward morphological classification is what inspired the Swedish zoologist Carl Linnaeus to create the naming convention that biologists still use to categorize every known species.

But it's also not hard to imagine a different system of classifying living things based on what they do rather than what they look like. Bees, birds, and bats can fly. Their wings are rigged up in very different fashions, but each of these animals, and related groups of animals, has analogous structures that address a similar problem. They have to get around to eat and mate and reproduce. They all converged on the same or a similar solution to the problem of flying. In biology, we call this convergent evolution, and we say their wings are analogous structures. Look closely, though, and you'll see that they are structurally very different, and they came into being at much different stages in the history of life on Earth.

Making sense of evolution really requires both ways of analyzing and classifying living things. If their physical configurations are different but their functions are similar or even exactly the same, those species are only distantly related. If the configurations of organs and bones are nearly the same, they are closely related even if their shapes are rather different. Genetic analysis is another way to test the degree of kinship between two species, regardless of how similar or different they may appear.

When it comes to flying, every creature faces the same challenges of physics, the same equations of motion and energy. To fly, we have to have air moving downward with enough momentum to support the weight of the flying object, be it a hawk, a yellowjacket wasp, an F-18 Hornet fighter plane, a flying fish, or a vampire bat. Watch a bird, the bigger the better. Their wings move down and back. If you've ever learned to swim the butterfly stroke, it's a similar motion. Instead of scooping up water and shoving it behind them, birds scoop air and force it down and behind them. This gives the animal lift and propulsion. In physics, anything that flows is considered a fluid. Thus water and maple syrup are fluids, but so is air. There's a trick in flying that you're aware of unconsciously or consciously.

When a wing or planar (flat) surface is moved through a fluid, it can develop lift or an upward force just by tipping it. It gets lift when it's given a so-called angle of attack, so long as it's moving. So once a falcon is airborne, it can develop lift by flying into the wind. It can climb using the energy of the wind and the angle of attack of its wings. An important consideration for any aircraft, be it nature-made or human-made, is that it be lightweight. Birds keep their weight down by having hollow bones and most of their wing area comprised of feathers. Made of a material very similar to your fingernails, feathers are remarkably strong for their weight, and they are grown so that they interlock and form a very nearly continuous surface. With the muscles below the attach points of their feathers, birds can pull the feathers down tight so that little air can pass between each feather. Or, birds can splay their feathers out like propeller blades or a fan. Each wingtip feather can provide some forward motion, while the inboard feathers provide lift. Birds can control the configuration of each feather and achieve remarkable efficiency.

Bee wings don't work this way, at least not exactly this way. Instead of having several feathers along a wing, each of which can be twisted to present a different angle to the air, bees have four wings. If you look closely, you can see the two pair. Furthermore, as soon as a bee gets in flight, the wings on each side of its body hook together with tiny hooks (hamuli, Latin for "hooks") along the leading edge of each rear wing to form essentially one wing on each side of her body. (Most of the bees we see are females.) Bee wings are flexible. With muscles that take up a great deal of the space in a bee's thorax, bees push their wings down and back, not unlike a bird's motion, but then they get wacky.

High-speed photography has revealed that bees turn their wings all the way over as they pull them from the end of their back-and-down stroke forward to start the next wing beat. The motion is hard to believe at first. Try straightening your arm and fingers out to your side. Reach forward with your palm at an angle to the floor or ground, imagine pushing air back and down. When your straight arm is behind you, take it up and twist it, so that the back of your hand is facing the floor and your palm is facing the ceiling or sky. With it still twisted, take it back to your starting position. Now, do that 230 times a second—not per minute, per second. It's astonishing. Bees get a component of lift or upward momentum on both their back or downstroke and their forward or upstroke. It's crazy in its way. They can pull this off—or actually pull this down, and back, and up, down, and back—because their wings are in sockets that let them twist in this—by human arm standards—exotic fashion. No wonder "Ripley's Believe It or Not" asserted that bees defy the laws of aerodynamics (this book's first paragraph). In the 1960s, bee flight was not yet understood.

Bats are a bit like birds and a bit like bees, wingwise. Like birds, bats cannot turn their wings completely upside down. Like bees after their fore and aft wings are hooked together, a bat's wing is one, albeit flexible, membrane. Bat, bird, and bee wings have this in common, though, the surface they present to the air is curved and flexible. Birds achieve this because each feather is able to move almost independently. Bees' wings are flexible membranes stretched between fluid-filled veins. Bat wings are bat skin stretched over bones. Each wing system scoops air and tosses it aft and down as it flaps. In evolutionary biology we say these wings are analogous. Now, the word *analogous* is regular enough, but here in this discipline it's used in a specific way to describe a specific type of relationship. Bees, birds, and bats use flexing wings to fly. But the wings came into existence by different routes over the course of evolution. They all use the same physics, but they are of very different configurations.

Look inside a bat's wing and you'll see it looks nothing like a fly wing. What it does look like is the bones in your arm. You have a humerus, radius, ulna, five carpals, five metacarpals, and fifteen phalanges. So do bats. And, get this, so do birds. This was one of Darwin's most important observations and insights. These bones are all in the same places in the arm and in the wings. It's just that each bone appears to be stretched or smooshed to fit in with the other bones to form an arm or a wing. In evolutionary biology, we call these structures and configurations homologous. They have the same shape, but they don't quite do the same things. You and I can't fly. Bats cannot play the piano. Birds can sing, but they can't hold a drumstick. (Uh, sorry . . .) I drew a sketch; it's on the next page.

This business of homology is one of the absolutely most compelling indicators of the process of evolution. Just by looking at our bones, you can tell that we *must* have something in common with bats and birds, and even pterosaurs, the flying reptiles that lived at the same time as the dinosaurs. The configuration of their bones is much like ours. It was a wild time back then: With more oxygen to power metabolisms, pterosaur wings were three times larger than those of the biggest living birds. They were like flying dragons, and yet also a bit like us. We are also a little like bees, but much less so. We have a central nerve running front to back. We have a mouth and an anus. We each have hearts. But otherwise, we're not so much alike. Six legs? With wings? Not my style, Ms. Bee. Fingers and toes? Well, sure Mr. Bat. Bees have analogs with bats. Bats have analogs and homologs with birds—and with us. It's wild. It's evolution.

Both analogy and homology are fascinating and keys to understanding where we all came from. Consider a fish and a dolphin. Fish

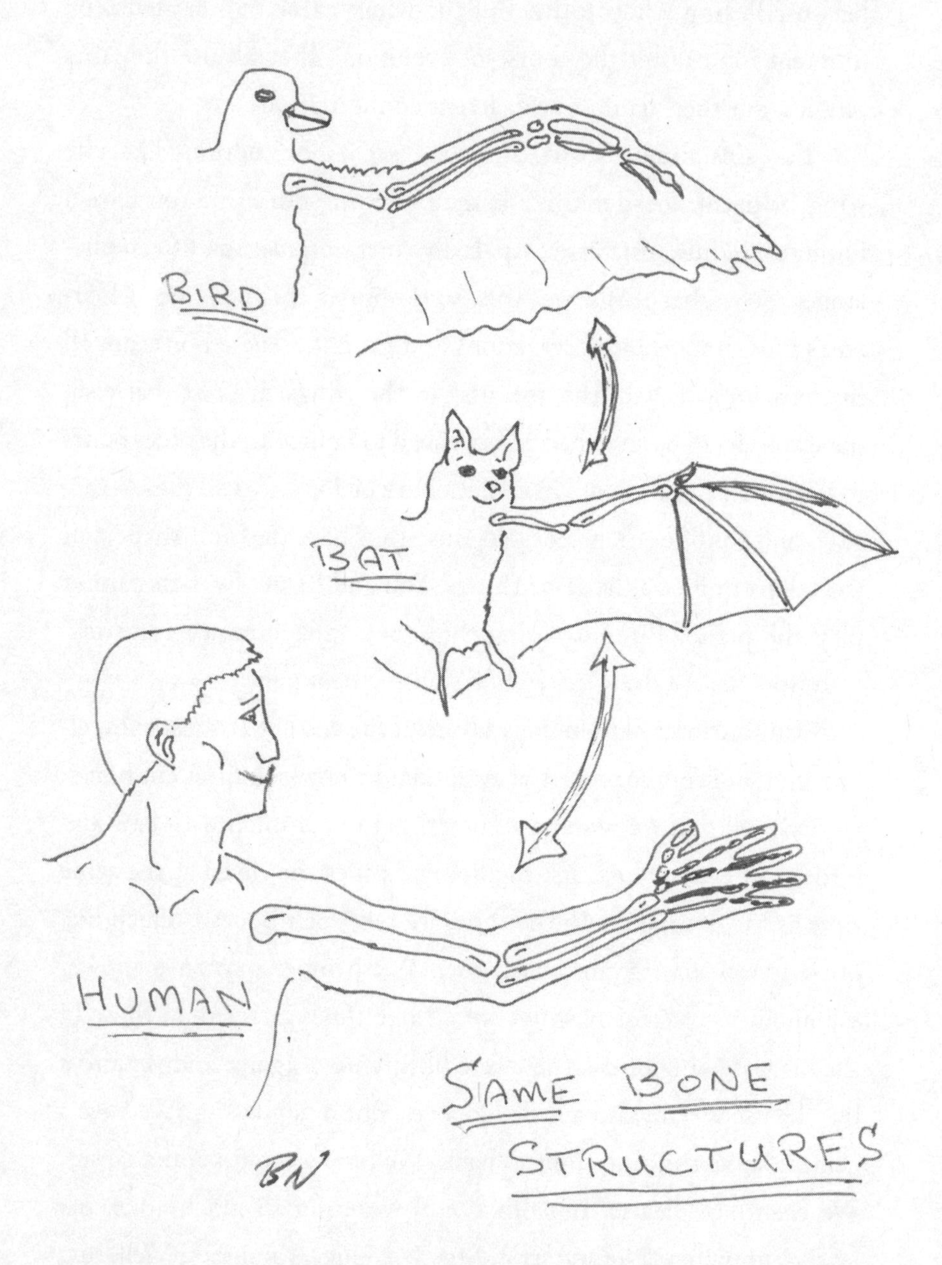

breathe the oxygen that's dissolved in water. Gasses can dissolve in liquids, like bubbles in beer. They stay in the liquid, "in solution," until the top is popped. A fish settles for whatever temperature the water around it offers. It runs its metabolism at a speed that the water allows. We say fish are cold-blooded, or exothermic, meaning "temperature from the outside." A marine mammal like a dolphin is warm-blooded; it has systems to keep itself at almost exactly the same temperature all the time, just like us. It uses calories to maintain that to be sure, but it can metabolize food efficiently, because its digestive chemical reactions are in a warm place. We say it is endothermic ("temperature from the inside"), just like us.

But check them out. Whether they're exo- or endo-, whether they have gills or lungs, they have about the same shape. They have to in order to slip through water with any reasonable efficiency. You might observe that fish tail fins are configured up and down, while whale tail flukes are set up side to side. You might think that they're quite different. Well functionally, not so much. Compare the flounder. It hatches ready to swim with its tail or caudal fin oriented up and down, vertically. But as a flounder matures, it rolls over and can lie flat on the ocean bottom. Its tail fin, along with its entire body, becomes horizontal. A flounder can swim without dragging its tail along the bottom. When we examine the ancestors of whales, on the other hand, we know they left the land and began to swim in the shallows. It's reasonable that horizontal flukes were better for shallow seas—no thwapping the sand as you swim. Later, as their descendants had success hunting in the open ocean, there was no need to twist their tails back. Horizontal flukes propel a whale well enough.

Many times while scuba diving or snorkeling, I've experimented with flipping both flippers together in whalelike fashion. You feel like you're moving a bit faster on the downstroke, but I sense an inefficiency on the upstroke. Whales and porpoises don't seem to have this problem. They aren't flipping with their leg bones the way I try; they flip with their spines. Since my spine is so much shorter than a whale's relative to the length of my body, I can't swim with the same oomph. I don't let it get me down. Even though we have perfectly homologous bones, I'm pretty sure I could beat a killer whale on a morning jog. Although if there were a mile-long pool next to a mile-long running track, the orca would win flukes down.

We can take this thought a step farther . . . or rather, a few steps backward. When we examine the shapes of extinct marine reptiles like ichthyosaurs ("fish lizard," roughly contemporaries of the dinosaurs), and even more ancient fish like *Entelognathus* ("full jaw fish," lived about 400 million years ago), we see the same streamlined hydrodynamic shapes we find in modern sharks, tuna, and orcas. To swim in the sea, you've got to keep things smooth. By the way, swimming in the sea is different from shooting like a rocket. Rockets are narrow-tipped and usually offer a straight tube from the bottom of the nose cone or fairing all the way aft to the tail (empennage). Thick up front and tapering to the tail is a streamlining feature of swimming and flying, when you're moving a lot slower than a rocket. So, fish and airplane wings are thick in the front and taper slowly toward the back.

Flying fish jump out of the water and fly, apparently to escape predators. I mean if you're a self-respecting tuna or mackerel, you eat fish. That's your business. You're swimming along and you see a fish that seems perfect for a meal. You approach at high speed and attack—jaws agape. And then, your prey does a few strokes with its fins, breaks the surface, and disappears. How frustrating for the tuna? Flying fish have been observed to link consecutive glides up to four hundred meters. If you're a mackerel, or a human out in a boat fishing, having the fish you're trying to catch move from where you are to a spot suddenly four football fields away in just a few seconds—well, it can be discouraging. Flying fish are found around the world at tropical latitudes. Why not? Who could catch them? Their regular fish fins are shaped to enable them to impart momentum to both fluids: seawater and air. They can slip through the sea and produce lift enough to glide through the air. Their fins are analogs of bird wings and homologous to the fishy predators they fly to avoid.

Analog structures in living things develop as different organisms make their way in their environment generation to generation. This tendency applies to plants as well as to animals. The leaves of trees and the leaves of sea plants (seaweeds) are an example: similar forms, arrived at completely independently. It's a universal aspect of evolution's adaptive imperative: Fit in or die.

As one embraces the processes that enable, indeed compel, one structure to be ever so slightly changed as its descendant bearer fits in with her or his environment just a little better than his ancestors, one can go again to my thought problem about the engineer trying to transform a bicycle into roller luggage. Each has two wheels. Each has a place to grip. If you are the craftsperson constrained by the rules of evolution, you'll have to come up with homologous structures from one design iteration to the next. The wheels will have to move from a side-by-side setup to a fore-and-aft configuration. And in each intermediate step, the whole thing has to function well enough to make it to its next generation. This pattern is revealed in the fossil record.

Life in all its forms must contend with the straightforward, inexorable rules of classical physics—of energy and motion. Sure, all of the organisms in our world are a result of the subtle and surprising chemistry that is ultimately a consequence of quantum mechanics and the interaction of particles smaller than atoms. Nevertheless, swimming, flying, pushing roots down through soil, drifting in the sea, etc., are all classical physics effects that are nevertheless every bit as astonishing as wondrous modern discoveries like the Higgs boson or the accelerating expansion of the universe. The laws of classical physics are sufficient to drive all of the convergent evolution and the analogous and homologous structures I've been discussing here.

What I find so compelling about evolution, convergent evolution especially, is that it is clearly a fundamental law of nature, like the laws of gravitation, electromagnetism, and heat transfer that shape our world. And yet it is much more personal than those other laws, because we are a direct consequence of it. Stranger still, we can understand it: nature comprehending itself from within.

20

WHAT GOOD IS HALF A WING?

A recurring theme among the skeptics of Darwinian evolution is that creatures are so perfectly suited to their tasks that they could not have become that way through blind natural processes. I hear it all the time. The exquisite wings of birds and bees, to take but two examples, are such marvels of engineering that they must have been designed by a deliberate creator. This line of thinking carries with it the mistaken idea that every biological structure makes sense only in its current form. If a hawk's wing is sheer perfection, then it stands to reason that it could not have evolved this way through incremental steps; otherwise, history would be littered with deeply flawed, incomplete versions of what the hawk is today. Often, skeptical creationists put it this way: What good is half a wing?

Like many popular criticisms of evolutionary theory, this one makes some intuitive sense, but only until you start looking at how the natural world really works. I am happy to address the question of what good is half a wing—or half an eye, or half a heart. Now that we've thought about good-enough design and convergent evolution and beneficial adaptions, we are ready to address this question, too. Please join me and take a look at Exhibit A, more specifically, exhibit *Archaeopteryx*, the amazing fossilized animal that resembles a bird and land-crawling reptile at the same time. The first specimen was discovered in 1860, just a year and half after Darwin published *On the Origin of Species*.

The most striking thing about *Archaeopteryx* is that he or she has feathers. The fossil is so beautifully preserved that you can clearly see their outlines. What did this animal do with them? Well, following the half a wing argument, the feathers must have been there so that *Archaeopteryx* could fly. Sure enough, when researchers look closely at *Archaeopteryx*'s fossil feather features they see the same kind of feather sockets or quill knobs that we see in modern birds.

In science, a hypothesis should not only explain the evidence we have found, it should also make predictions about things not yet discovered. Knowing that *Archaeopteryx* had feathers, evolutionary biologists predicted that there should be other transitional forms between bird and reptile, and there should be transitional forms of feathers and wings. Something extraordinary happened over the past two decades: People digging in previously unstudied fossil fields in China found the remains of feathered dinosaurs. Not just one or two dinosaurs, but many different species. Furthermore, the evidence now shows that a lot of other, familiar dinosaurs had feathers as well. Perhaps they all did; the feathers just weren't preserved well enough to show it. This is or was true even for land-dwelling predators like the velociraptors that starred in the movie *Jurassic Park*. (What's that you say? You don't remember seeing feathers in the movie? That's because it was made before paleontologists discovered those feathered fossils. Science is inherently a work in progress.)

Velociraptors clearly did not fly. These dinosaurs had big thick leg muscles and arms too small for winging it. Nevertheless they had feathers; as with the *Archaeopteryx* fossils, we can see their feather anchor points, the quill knobs. You have to figure velociraptors must have had feathers for reasons other than flying.

It seems to me the most likely purpose of the feathers was either to keep the animal warm or to make it hot . . . and by hot, I mean sexually attractive, to show that it was well suited to reproduction. In fact, those are the two leading theories among experts in the field. Hey, mulling it over some more, maybe their feathers performed both functions. They kept these animals warm and they helped a velociraptor show off, like a stylish winter coat.

But what about our scientifically beloved *Archaeopteryx*? What of her or his feathers? Were they airworthy? The answer seems to be clearly: maybe. It's intriguing to say the least. If you've ever recovered a bird feather, you can see that there is a central quill, and on either side is what we call the "barbs." Those are the feathery parts of a feather. Furthermore the barbs are held to each other by other structures we call barbules (little beards) and hooklets (little hooks). The whole thing becomes quite rigid considering how crazy lightweight it is.

Among the remarkable things about a feather is that its shaft is hollow yet sturdily structured at the same time. If you've ever seen a newborn baby chick, you may have noticed that he or she has thin feathers; they look almost like individual long hairs. They also have scales, not too different from an alligator's or snake's scales around their cute little happy chick feet. Feathers, scales, and hair all start from the same kinds of cells. Each of us, snakes, birds, and people, has the ability to produce structures made of keratin, the natural plastic that is a snake's scales, a bird's feathers, and your hair and fingernails.

Researchers have looked ever so closely at the fossilized remains of *Archaeopteryx*. The way their bones are arranged makes most scientists think that these animals could not raise their wings above their heads. All the birds that you and I know can. Big upstrokes help modern birds push enough air down and back quickly enough to produce lift. It's not clear that *Archaeopteryx* could fly with this kind of modern motion. However, and this for me is huge, the feathers found with and near the *Archaeopteryx* fossils were not symmetrical, which implies that these animals could fly, at least to some extent.

If you fly airplanes, this next bit is obvious. If not, next time you're around airplanes, notice that the wing is thicker in the front than the back. As a very good first approximation, wings are thickest at a point that's about a quarter of the distance from front to back, from leading edge to trailing edge. The leading-to-trailing dimension on a wing is called its chord (like a line segment in a circle). The thickest part of the wing is at the quarter-chord point. That's where we humans run the biggest support beam or spar. It goes lengthwise through the wing (crosswise to the body of the plane). Well, the feathers birds use for flying are about the same. The shaft or quill of a feather used for flight runs along a feather's quarter-chord point. The feathers of the *Archaeopteryx* run the same way.

On modern birds, and on ancient fossil birds, the tail feathers are *not* asymmetrical. Instead, they are nearly the same left to right. Those feathers are not subject to the forces or wing loading of a bird's flight feathers. And, wait . . . wait . . . there's a little more. Modern birds have extra feathers called coverts on top of their primary feathers. On modern flying birds, the coverts are used to smooth the airflow over a bird's wings. They work exactly like the fairings we put on our human-designed airplanes. Look under the wing of a modern jetliner, you'll see the long so-called canoe fairings that keep the air going smoothly around the mechanisms that control high-lift surfaces like flaps. The fairings add weight to the airplane, but they're worth it in cutting drag. The covert feathers on a bird add a little weight as well. They also take energy for the bird to grow and replace as feathers get beaten up in use. It's akin to your fingernails growing continuously, because you wear them down. (Try putting adhesive tape over your nails for a couple hours. You'll see how much we use our nails. It's surprising.)

Empennage refers to the tail feathers or fletching of an arrow, the tail of a bird, or the tail of an airplane. Note the Latin root *penna*, which means "feather." For centuries, humans used quills for pens. While we're on the subject of empennages, note well that a peacock's main tail feathers are much like other birds' tail feathers. The wild and striking plumage that we see displayed in courtship is made entirely of elongated, dressed-up coverts, big little feathers. Just like all other birds that fly, peacocks don't subject their coverts to high lift loads, even though their coverts are huge.

In a remarkable investigation, scientists analyzed the fossils of *Archaeopteryx* feathers and the fossil feathers of related species, with an exquisite X-ray system. They determined that there is over an 80 percent chance that the feathers were dark or black, like a raven or crow. If you've ever thrown a Frisbee, you may have noticed that darker-colored disks are stiffer. The same is true for food storage containers, the more opaque, the harder and stiffer. The plastic's so-called pigment loading affects its stiffness. So it is generally with keratin, the stuff of feathers. Dark feathers would be stiffer and perhaps better suited for flying.

In 2013, an eleventh specimen of *Archaeopteryx* was unearthed. Apparently, *Archaeopteryx* evolved around the same time as several other feathered reptiles. Analysis of *Archaeopteryx* "trouser feathers" around his or her ankles indicates that he or she probably could fly at least to a limited extent. Even if the feathers weren't suitable for full-time flying, other contemporary species did have airworthy feathers. The crime scene is ancient, and the jury is still out.

I admit it; I'm fascinated with living things that can fly. Perhaps I'm just envious. It strikes me as quite a clever strategy. If you can get airborne you can really cover some ground, or some ocean. You can avoid predators, find food, and shop for a place to roost all without getting your feet wet or dusty. But flight for any organism or human-built machine is a complicated business. In engineering we say that if you have a big enough engine, you can make anything fly. In many ways, the more difficult problem is steering. Imagine a car that you couldn't steer. You wouldn't need a warranty, because it would be guaranteed to crash. Flying takes continual, accurate steering inputs in three axes: rolling, turning, and up and down-ing (roll, yaw, and pitch). Without control in flight, a bird would be a dead duck. So did *Archaeopteryx* have the wits to do it?

Careful study of the skull of an *Archaeopteryx*, specifically her or his brain case, indicates that she or he not only had feathers and wings sufficient for flight, she had a brain big enough to fly. Very nearly all of our planes have horizontal stabilizers or tail-planes and vertical tails, the part that sticks straight up and has the moveable rudder incorporated. But now consider the B-2 bomber airplane; it has no vertical tail, no tail at all. Military tacticians wanted to eliminate the vertical tail, because radar would bounce off of it easily. Any plane with a vertical tail is a great deal easier to detect than planes that would somehow not need one. Keep in mind that these tailpieces keep the plane going where we want it to go, exactly like the fletching on the bowstring end of an arrow, the arrow's empennage.

It was quite a deal in my little world of military airplane design (I had security clearance for a while, and so on) to design a plane with no tail sticking up. Obviously for a bird to go without one is easy, a day in the office . . . or nest for them. Humans, at great expense, designed a flyable plane without a vertical tail, and we did so perhaps 150 million years after birds did. It took us years of research and the development of flight-control computers that could work fast enough to continually adjust the other control surfaces like ailerons inboard and outboard to pull it off. Perhaps *Archaeopteryxes* had no such trouble, because they had the feathers and the brains for it.

With all this, it's reasonable to infer that *Archaeopteryx* could fly a little bit, or perhaps well enough to ride the wind and columns of rising hot air (thermals) the way so many modern birds do. It's fun to consider. Even if they really couldn't lift their wings in modern bird fashion, even if their feathers weren't really long or strong enough to spread their weight out sufficiently, even if their brains couldn't pilot them over a warm sea or through a thick forest, well, they still must have *almost* flown.

However you cut it, *Archaeopteryx* basically had half a wing. *Velociraptor* had something more like one quarter of a wing, or one eighth of a wing. And you know what? Those intermediate steps worked just fine; they were just adapted to slightly different functions than the wing of a hawk. Perhaps *Archaeopteryx* could glide from limb to limb. Maybe he or she had the ability to jump off a high branch or cliff and land slowly, without breaking a leg or neck. Maybe they could just barely fly. If that was the case, it would be as though they had just half a wing. What if all those feathers did nothing more than keep these animals warm? What if all my musings about their aerodynamic performance was just that—thinking out loud without a strong basis for doing so? Then what if the feathers were just adornment for making time with others of their kind? That would be good enough, in the evolutionary sense, and very useful. Half a wing was all an *Archaeopteryx* needed to survive.

Archaeopteryx was, in our modern evolutionary way of thinking, a transitional animal. Its fossils are transitions from feathered dinosaurs to birds. And, if you haven't heard, our birds flying around today are indeed direct descendants of ancient dinosaurs. We see it in their feet, their fossils, and their feathers. Darwin speculated quite reasonably about this relationship long before X-ray analysis was available or quarter-chord points were postulated. He was a thoughtful young man.

But here's a crucial point: At the time when *Archaeopteryx* lived, it wasn't a transitional *anything*, or a half anything. It was a creature well adapted to its environment. To a human designer, *Archaeopteryx* looks perhaps like an incomplete version of what we expect a bird to look like, because our expectations are set by the birds of today. In its time, *Archaeopteryx* was a perfectly complete competitor in its ecosystem.

If these ancient protobirds were like our modern birds, and they used feathers to keep warm just like modern birds and we do, it is quite reasonable to presume that the ancient dinosaurs were able to generate their own heat, just like us. How about that? Instead of slow-moving, sun-basking snakelike animals, they were quick like a hummingbird or a hawk. The scary-fast dinosaurs of the *Jurassic Park* movies were inspired by these fossil discoveries. And the science keeps getting richer. Some of the newest studies suggest the dinosaurs were not entirely warm-blooded but weren't cold-blooded either. They may have been something in between: Half-warm blooded, which (like half a wing) evidently suited them just fine for many millions of years. I'll use a new term: *mesothermic*, meaning "heat from somewhere in between in and out."

Feathered dinosaurs are one among a series of recent discoveries that are filling in the story of evolution by illustrating the transitions between one kind of organism and another. The exact transitional moments may be almost impossible to find (a characteristic feature of punctuated equilibrium), but in many cases it's possible to track down the more general intermediate forms—with persistence and some smart strategy.

For many years, evolutionary scientists were frustrated that they could find no transitional animal between fish and land animals, like lizards, crocodiles, and alligators. That is, until they got to thinking about it. They figured such a creature would live in a swamp or a marsh and he or she would have had to be living about 375 million years ago. When people discovered a fossilized swamp that tectonic plate movement had carried north to what is now eastern Canada, they went looking. That is how Neil Shubin managed to find the remarkable *Tiktaalik*, a fish with transitional features between fins and legs. In the context of our current discussion, you might say *Tiktaalik* had half-limbs. But from the creature's perspective, there was no such thing as a half anything. What it had was something that enabled it to crawl onto land, to escape predators, perhaps, or to get a better look at potential prey.

I cannot emphasize enough the significance of this discovery. What we want as scientists, and as non-geek people who use the scientific method, is prediction. We want to come up with theories that allow us to make predictions about the future. It's in our nature. Our ancestors, who did not bother to make predictions about the future, no doubt got very quickly outcompeted by other ancestors who could predict seasons, the movements of herds of prey, and the growth of food plants. Scientists predicted that an animal like *Tiktaalik* would be found, and they found it. I drove this point home to my creationist adversary in the Kentucky debate. Creationism, unlike science, can predict nothing. Along with it being obviously wrong, it is obviously not useful. And of course, an appendage that serves as both foot and fin was no doubt extremely useful to animals like *Tiktaalik*. It's how they got around, hunted, escaped predators, and reproduced.

The efficacy of half a wing is an old question that has a clear, compelling answer in the context of evolution. Here's another fantastic example of how evolution produces intermediate forms. In the

1990s, a group of fossil hunters found the remains of a whale that once walked. No kidding. *Ambulocetus* (walking whale) fossils were discovered in what is now Pakistan. It is an animal that has whalelike flippers, and feet with toes.

Ambulocetus must have roamed the shallows, where, judging by its teeth, it ate other animals. Through chemical analysis of those teeth, investigators have determined that *Ambulocetus* was able to make the transition from salty water to freshwater. They lived in the estuaries, where rivers meet the seas. They probably had no trouble finding food in these nominally productive areas. And they had hair, enough of it to be apparent in their fossils. These animals evolved into our modern whales. They had both half a flipper and half a hoof. You can be certain that they made outstanding use of both. *Ambulocetus* lived in large enough numbers, and retained reproductive capacity long enough, for us to find their fossils 50 million years later.

Half a wing, half a foot, or half a fin, they were well suited enough to outfit their descendants who are flying, walking, and swimming among us to this day. Each feature had to work well enough in its time. And each one did—well enough.

21

HUMAN BODIES ARE WALKING, TALKING, AND GOOD-ENOUGH

You don't have to dig into the past to see examples of good-enough design. You don't even need to visit the zoo. May I respectfully suggest you just walk up to a mirror? You and I, and our parents, and their parents, and your kids are the walking, talking, dancing (sometimes) result of evolution's principle of "good enough." You and I will wear out. You might already have aches, pains, eyeglasses, and dental fillings. But your generation, no matter which one it is, was good enough to make it this far. This is another consequence of being shaped by natural selection. In our evolutionary world, good enough is as good as it gets. There is no reason for nature to work any other way. There's no evolutionary pressure to produce designs that are better than they need to be.

Every single feature of an organism takes energy to produce. Your hands, your eyes, your brain—all of these require chemical energy from somewhere, and the plan or design for them came or comes from your DNA. No entity at the top of the Human Being Design Shop is anticipating what feature we will need in the future. Whichever organisms, with their (our) randomly generated and sexually selected mutations, happen to come out better suited to the world they're born into—those are the ones, like you, that have a shot at reproducing. Either your features, such as your eye color, fingernail thickness, elbow joints, and emotional disposition, enable you to live long enough to reproduce, or they don't. Nature's good designs outcompete her not-so-good designs.

This fundamental feature of natural selection explains something we all worry about. Or if we're not exactly worried about it, we are at least keenly aware of it: We are all going to die. I know, it sucks, but it is the way of our world. From an evolutionary point of view, or from the point of view of your genes (if they have points of view, as such), what difference does it make? That you have this great big brain and you can worry about it is, in a charming sense, your problem. Evolution, with its 16 million species of animals, plants, microbes, and viruses, continues whether or not you or I express concern. All of us living things, from sea jellies to zebras, have to play the genetic hand we're dealt.

What really puts this in perspective for me is the fictional world of superheroes. Everyone is aware of characters like Superman, Spider-Man, and Wolverine. They have superpowers. O, would but that we could fly! How cool would that be? Or what if we had superhuman strength? Wouldn't it be nice to be able to outwit every villain you can think of? For me, these comic book heroes help us imagine what we'd change about our biological selves. They are also a means to nurture logical thinking. I'm not kidding. What if you, like Green Lantern, could move objects all over the place with your ring . . . unless the object is yellow? On one level it's silly. On another level, it's a mental exercise that helps you notice a detail in the world around you that many of us might otherwise overlook.

Batman is important in this discussion. Wait, hear me out. He doesn't have superpowers. He's just a brilliant man and a superb athlete, who never loses his cool, and is astonishingly wealthy. Oh man, I want to be that guy! But even Batman is replete with human features that a thoughtful designer—an intelligent designer, in the term favored by some creationists—would improve.

One of the most obvious human design puzzles is that our waste disposal plumbing is immediately adjacent to our reproductive and pleasure producing plumbing (even for Batman, I assume; though that information is not generally available). Your anus is right next to either your penis or your vagina. Would you have put the urethra right there in the middle of the whole business? If you were in charge, wouldn't you have separated those a bit? How hard could that be? (A car's air intake and its exhaust pipe are on opposite ends of the car, aren't they?) Seems like a simple problem to correct. Speaking of air intakes: why is yours right next to, actually right on top of, your fuel input system? Your windpipe (trachea) is right next to your food tube (esophagus), making it easy to choke. What's up with that? Couldn't that have been improved?

Why does anyone need to wear glasses with corrective lenses? Wouldn't you, as the designer, have just specified perfect vision for everyone? Why do so many of us get sunburn? Why isn't our skin more sensitive and tougher all at the same time? Same with hips, knees, and anterior cruciate ligaments: Why not make them a lot more durable? We wouldn't be having all these replacement and repair surgeries.

Next time you look an octopus in the eye, respect her or him, because her or his eye is a better design than yours. We have a blind spot near the middle of our retinas where our optic nerves connect. Your brain has to fill in the missing piece of the picture so you don't notice it. Moreover, the human eye's light-sensing cells are tucked behind other layers of tissue, which creates a slight distortion. That's not an optimal optical arrangement, but that's how we evolved. Octopus eyes don't have either of these problems. The eyes of octopuses and humans came to be by way of different evolutionary paths; and here we all are.

Every third pitcher in Major League Baseball in the United States has what has come to be called "Tommy John" surgery (after the first player to undergo the procedure). The ulnar collateral ligament, which runs around everyone's elbow, wears out. Surgeons harvest a ligament from somewhere else in your (a pitcher's) body, drill little holes in your bones, and tie the replacement part into your arm. The procedures work very well. Modern surgeons literally can't count how many of these operations they've done. If you were designing the elbows, wouldn't you make the elbow ligaments a good deal better or tougher? Or, wouldn't you make it so humans didn't want to play baseball? Perhaps better still, make it so that our brains were able to stop us from continuing to play before our elbow ligaments were worn out? Nope. Pitchers play the arm they're dealt. Nowadays, a third of them rely on other humans to fix their overuse problems.

Now that I think on it, I believe the brain is the biggest problem here. First of all, why do we need to sleep? For cryin' out loud (as babies are wont to do), couldn't we design a brain that can just keep going 24-7? Our computers run day and night with little difficulty, and they have not had the better part of a billion years to get things right. Why not us? Why do we get confused (in my case, more confused)? Why don't we have brains that just figure everything out, just like that? How hard could it be? If a few of us can figure out how to do arithmetic problems in a flash, why not everyone? Why can't we all be born able to work calculus problems? With the right designer in charge we would all come fully loaded, as we say at the car dealership. But we don't, and evolution is why.

Our bodies, like those of every other living thing on Earth, reflect the physiology of successive generations of ancestors who were built well enough to reproduce. That's all they had to be: good enough. People who have or had trouble accepting the process of evolution, often point (but not poke) at a human eye and marvel at its construction. They express stern skepticism that such an efficacious and wonderful structure could come to be without an all-powerful designer running the show. Really, this is another variant on the "half a wing" argument, and it's just as easy to refute.

If you take a little time with it, you'll find countless examples in nature of organisms that have light-sensitive cells. It's easy enough for us to imagine cells that sense heat. Light and heat are the same energy, just at different wavelengths and often at different intensities. In nature we find light-sensitive cells on a flatworm, say. We find light-sensitive cells lining a hollow or pit on the shell of a mollusk; eyes that work well enough for them. We find light-sensitive cells in nautilus eyes that are illuminated through a very small opening, just like the pinhole camera my father rigged up while interred in a prisoner-of-war camp. You find spider eye lenses that give them a sense of where the light is coming from. You find repeated-pattern compound eye lenses in insects, and so on.

Our eyes are derived from millions of years of trial and elimination of error. You'll find organisms with transparent cells. In other words, there are a great many examples of the intermediate steps required to get from a flatworm's "light patches" to a giant squid's manhole-cover-sized eyeball, to the eyes of a bald eagle that can resolve images about eight times better than we can. It's as though they have eight times the megapixels in their smartphone cameras. The intermediate steps are still extant. Each of the steps I mentioned exists in nature today, because each of these eyes is good enough to keep each animal and its genes in the game, the game of life.

Speaking of being in the game of life, who, in your estimation, is the most dangerous animal on Earth? Who is the baddest badass? It's you, of course. It's us. We are the dominant entity around these parts. If you're a cow, humans will breed you, milk you, kill you, and eat you. If you're a mouse, you can run around underfoot, but humans will exterminate you, if you get in our way. Heck, you can be a whale in the middle of the ocean, and humans will build ships big enough to sail on the open sea, hunt you down, and kill you. Humans are serious business because we have big brains (even if we don't all use them fully—again I'm reminded of my old boss). So let me take you down a peg by pointing out that the human brain, the exalted thing that elevates us above all other species, is just one more example of evolution's good-enough standard.

Our brains are radically bigger relative to our body weight than the brain of a horse, for example. I believe that's part of the reason horses can shy and act a little wild and, by human standards, irrationally and dangerously. Our brains are enormous compared to a dog's, even a sympathetic, happy, smart dog such as yours no doubt is. Our brain-to-body-weight ratio is somewhat bigger than that of our genetic cousins the chimpanzees. But our brains are not especially much bigger in proportion to our body weight than that of our former neighbors the Neanderthals. We outcompeted them and perhaps two dozen other almost-modern-human, or "hominid," ancestors. We are just the next step. Our brains, compared with our bodies, are just a little bigger than the other guy's or gal's. Nor are our brains unique in the way they work. Have you ever watched a dog when he or she is asleep yet still a little active? They move and shake very much the way we do. You don't have to be a neuroscientist to figure out that our best friend is dreaming. Their brains are hooked up just like ours.

We are just the latest version of a brain good enough to conduct our lives in competition with so many other organisms in our ecosystem. As a consequence, we can communicate in ways that, at least from our point of view, seem to be far more complex than our next of genetic kin, the chimpanzees, bonobos, and gorillas. We can write books, and thank our readers for reading them. Thank you. We can understand patterns in nature, record them, and make predictions based on those patterns. We can discover calculus, write plays, develop movie projectors, and make moving-picture stories about a future that may or may not come to pass. This very discussion about evolution would not be possible if we did not have the uncanny ability to step back and be aware of our situation here on Earth and in the cosmos. A better brain is just not the same thing as a perfect brain.

Although our brains enable us to envision a remarkable future, it's not clear that we'll be able to fly around the galaxy or universe, beam down to a planet, and chat it up with beings who speak English (for example). And who knows? Some future, far-more intelligent species may look back at us as the transitional form with "half a brain."

Bear with me for a little more delightful evidence of good-enough design in our brains. Studies have shown that not only does your brain direct your movements, your movements can affect what you think and how you feel. When test subjects are induced to move toward a stranger, the subjects tend to like the stranger more than when the subjects are induced to move away. This may be why shaking hands and bowing forward became traditions. By moving toward the other person, we feel, at some level, that we accept the other person. Psychologists and neuroscientists have coined the phrase, "embodied cognition." When your brain directs your body to do something, your body, in a sense, directs your brain to feel something. I mention this, because it is, for me, yet another sign that our brains are derived from our ancestors' brains. In comedy writing, we would call us a derivative bit. We're just like those other hominids only updated to be funny (or tall, or smart, or cute, or charming) today.

And the brain, like every part of the body, is mortal. Right now, all manner of research and medical work is being done to try to prolong our lives. This may work out for future generations, but I've got a feeling that even if people live to be two hundred years old, it won't make that much difference in the evolutionary scheme of things, because what matters is getting one's genes into the future. It's as true for you and me as it is for lobsters. A human's reproductive period is probably going to remain only about thirty years long, even if that person goes on to live and pay taxes for many more decades. That we age, that we wear out, is not of much consequence in the evolutionary scheme of things so long as we reproduce. Having thirty years of reproductive time was good enough for our ancestors, and for better or for worse it has to be good enough for us, too.

It is reasonable to me that the length of a human's natural life is also the result of natural trial and error. If we lived too short of a life, we would not come of age and be sexually able to make more of us. If we live too long, it is reasonable that it would take too much out of our immune systems to keep fighting the next generations of germs and parasites. Instead our seven, eight, or nine decades are the right amounts of time naturally to be productive without being a burden, either to our own immune systems or our tribe. It's a troubling thought or an empowering one, depending on how you look at it.

Suppose you knew that you were going to live to be two hundred. Would you hustle as much? Would you work to better your life? Would you bother to learn algebra, or would you assume that you could just learn it later? Would your reproductive time be any different? Would women be able to conceive children after the age of sixty? Would a man's sperm not degrade with age? Extending life might be fun, but it probably wouldn't make too much difference in our reproductive ability and our evolution in the long run.

All of us, everyone reading these words, have made it this far in life. None of us would be here if we weren't genetically good enough. That's a rather encouraging thought. We celebrate certain people's appearance or their wit, but we are all so much more alike than we are different. The proof is in the living: We all made it. No matter how ugly you think someone else is, he or she got here just like (as) you did. There's a lid for every pot, as the saying goes.

22

EVOLUTION IS WHY WE DON'T BELIEVE IN EVOLUTION

From time to time, I meet someone who will say something like, "I am not afraid of dying." I don't buy it. *Everyone* is afraid of dying. It's part of the instinct that helps us survive as a species. It's a crucial feature of human evolution. It's also, I strongly suspect, a crucial reason why so many people have trouble believing evolution is true. Life can be ironic like that.

It doesn't take much to trigger the fear of death. Just try to picture these scenarios. You're standing on a very high bridge, like we might have near the campus of Cornell University in New York State. The gorges there are 50 meters deep (about 150 feet). You are scared—because of your fear of something going catastrophically wrong, that you might lean to have a look, and lean too far, etc. Or, you're crossing the street, and a driver who's texting almost hits you. Perhaps he slams on his brakes at the last possible moment. The tires squeal, the horn blares, your heart almost stops. You are scared out of your mind because he almost killed you. We are all scared of things that go bump in the night, because it may be a sound made by something dangerous, e.g. a lion, tiger, or bear—that could kill you.

Our ancestors who did not have a fear of heights, who did not have a fear of eating something poisonous, who did not have a fear of venomous snakes and spiders, who were not afraid of drowning, well—they're dead. They did not have the instinct to keep themselves from getting killed. It's deep within us, and it had better be. I thought a lot about the fear of death after I debated creationist Ken Ham. There is a deep-seated reason why intelligent, sensible people suddenly recoil from objective evidence when the topic turns to evolution. I think the fear of death has a lot to do with it.

As you may know, a group of wags on the Internet issue what they call the Darwin Awards. These are largely just news stories about people who behave in extraordinarily stupid and dangerous ways. Many of the stories seem to be apocryphal, but they sure make for good reading. For example, a guy may find some old dynamite, bury it, and then tamp it down real well to make sure the soil fills in around the old sticks. Well, during the stamping and tamping, the pile explodes, and the man essentially vaporizes, leaving almost nothing for the police to examine. Or, there's a story about a guy, who tied jet-assisted takeoff (JATO) rocket tubes to the roof of his car. When the rockets were lit, the whole assembly became airborne and flew into a high mud cliff, flattening car and experimenter alike. Each of these people is given a Darwin Award, albeit posthumously. The moral is that they did not have sufficient fear of death, and it killed them.

There's a scientific message here, too. Evolution does not influence only physical attributes like height, number of knuckles, eye color, and earlobe shape; it also acts on emotions. What we feel is a result of evolution. This is true, no doubt, of your fear of death. It certainly seems to be true about our drive to reproduce. When we're thinking about sex, what we're really thinking about is what it would take to engage in activities (sex) that would lead to reproduction (babies). It goes on all day, all the time for most of us. I encourage you to place your own sex joke here, and notice how readily it came to mind. Sex is there, just below the surface, all the time!

If our genes drive us to remain alive and reproduce, and we are unconsciously driven all day and night by this impulse, what does that mean about the rest of creation—about all of nature? I like to compare us, or me at least, to dogs. I've watched them. They dream. They have fears. They certainly can be charming, and they can be quite difficult to be around, just like people. But, it's not clear to me that my dog friends get quite as caught up in the nature of our existence as my human colleagues and I do. I have seen dogs very afraid—of a mean or angry owner, or another, bigger dog. But I don't think I've ever seen a dog paralyzed by self-doubt, as *I* have been from time to time. I am open-minded, but I don't think dogs, even dogs I've spent a lot of time with discussing these things, ponder the origin of the universe and the fundamental meaning of life. Leastways, not the way you and I do.

Apparently, a consequence of having a human brain that can do all these other remarkable things like play the violin, invent a trick in calculus, or pole-vault over walls and thin striped bars, is being able to ponder our very existence. No other organism on Earth does that. Yes, yes, I understand that dolphins are very smart, but I don't think they build libraries or even contemplate such a project. Along with our special ability to think and reason, we just can't believe it's all going to end, that we are all going to die. But near as I can tell, everyone that's ever lived is either dead or is going to be. What about this house you designed? What about the poem you wrote? What about the feeling you had, when you fell in love? Those things are not going to go on without you; generally they'll disappear in a flicker. It is the miserable nature of our existence.

If you live to be eighty-two years old plus about seven weeks (it depends on leap years), you get thirty thousand days here on Earth. That's it! When I was a boy, thirty thousand of anything was hard to imagine. It sounded like a lot. I just think how many balsa wood airplanes and model rockets I could have flown and launched with thirty thousand dollars! Now, as an adult paying taxes and writing books, it hardly seems like a big number at all. Try this way of picturing a human lifespan. The National Football League's Dallas Cowboys' stadium holds 105,000 people. Now, imagine that you're watching life go by down on the field, and every day you watch that life go by from a different seat. You don't even get a third of the way around. Before you've settled into a third of the seats, you'd be dead. And, that's if you had a good run, eighty-two-plus years. Yikes!

Are we the only organisms around here that have this sense—a sense of doom? It sure seems like it. I spent a great deal of time in the Pacific Northwest watching and swimming with salmon. When it's time for them to head upstream to mate, they stop eating. They get their chemical energy, their food, by digesting themselves from the inside out. They essentially eat their own intestines on the way upstream to their deaths. Do salmon know that they are going to die as they head upstream to lay eggs or deposit sperm? Or, are they like a teenage boy driven by sex, sex, sex?—they can't think, or they actually *don't* think, of anything else, not even food. Or are salmon aware of their impending end of life, and are they sad? I must say it just doesn't seem like it. They swim; they mate; they die. It doesn't seem like there is a lot else going on with them.

Apparently, our closest relatives, the chimpanzees, suffer a few days of depression or mourning, when a member of their troupe or barrel dies. Do they perform rites associated with the death of a colleague or mate? If they do, I am skeptical that they take it as seriously as humans do.

But humans?! We go nuts. We construct all sorts of systems to reassure ourselves that there is more to it than swimming, mating, and dying. We deposit our writings and musings in libraries. We build statues to people whom we admire, so that at least in one sense these people live on, or their memory does. Heck we name buildings, highways, and mountains after certain people. We keep letters of deceased ancestors. We erect grave markers. All of these things, in some way, preserve a life, or at least a life's work.

It seems that we really are special, so why shouldn't there be something special waiting for us on the other side of Cowboys' Stadium? But when we're waning, when we're dying, it's not obvious that our minds are anything more than a product of an exquisitely complex system of chemical structures and chemical reactions. I remember well my grandmother talking with me about wildflowers. She had an extraordinary memory, and she had done quite a bit of artwork that included pastoral scenes of New England and a great many flowers with a naturalist's attention to detail. She spoke to me about pollen, pistils, stamen, and ova. I remember a few conversations about baseball. She listened to it on the radio, sometimes with remarkable attention to detail. She would comment on Don Mattingly's swing, even though she couldn't see it. She just thought about what she heard and where the ball went, and so on. Toward the end of her life, though, her amazing mind slipped away. She became incompetent with regard to baseball, wildflowers, the delicate control of her artist's pencils, and just about everything else.

I worked closely with Bruce Murray, a planetary scientist at the Jet Propulsion Lab (and founder of the Planetary Society) who had a tremendous influence on the United States' space program. He insisted that the early space probes should have cameras on them, and is largely responsible for the first pictures humans ever captured of the planet Mars. Toward the end of his life, Bruce could tell you a great deal about those heydays, but he could not remember whether or not he had had lunch, let alone what he had had for lunch. It's heartbreaking. At the Planetary Society we honor him by naming our collection of pictures and videos the Bruce Murray Space Image Library. His loss of memory is evidence of the chemical construction of our evolutionarily built, good-enough brains. Despite many human beliefs about an afterlife, it sure seems as though these remarkable people did not have their consciousness transported to some wonderful eternal place of rest and contemplation. Instead, it seemed as though they lost their faculties as certain systems in their bodies shut down.

Blame evolution for this unsettling conflict between the way the world works and the way we wish it would work. The fear of death, combined with a novel ability to envision the future, enabled humans to outcompete other species. But that combination also makes us unable to believe that what we see around us is all there is. Our brains got too big to think about the world any other way.

I guess the question of whether or not we're special is another way of asking, what's the point of our having big brains? Diligent paleontologists and archeologists have combed hills and valleys looking for earlier versions of us. Researchers have found dozens of bones and skulls that once belonged to our distant, distant relatives. As we study these other humanlike almost-people, we realize that they were almost like us. As we examine our recent humanlike ancestors like Neanderthals, or Cro-Magnon, or other humanlike ancestors, it becomes clear: If these guys and gals were dressed like us, you'd hardly notice them on a busy sidewalk. It just seems as though they were almost like us. Their brains were almost like ours. They almost embraced the same worldview that we have with the same suspicions or beliefs about a life after death.

It looks like we came out with the most versatile brain, which could throw, catch, and hit balls—and invent the rules of games involving balls, better than they could. It looks like these same brains gave us the ability to wonder about our place in the scheme of things, and that led us to science, and that led us to the discovery of evolution. We are products of evolution and as such, we can't believe it's all over, when it's all over.

It is a great irony of the evolution of the human brain: Our strength is also our weakness. Our asset is our liability. Through millennia of refinement, we have the ability to recognize patterns better than any other organism out there. Sure, monarch butterflies head south, when it's the right time of year. Deciduous trees shed leaves, as the days get shorter. They sprout new ones, when their internal chemical clock says it's time to. But no organism out there creates calendars with leap years. No organism out there launches rockets that go into space and back, boosted and directed by chemicals and physics. I can imagine, though, another tribe that could almost do these things. It was just that we recognize patterns a little better, so we beat them to the best food sources and shelters. And now, we're left with these brains that can make us crazy.

Which brings me back to Ken Ham and his belief, widely shared among his followers, that Earth is only 6,000 years old. When I debated Mr. Ham, I talked a lot about the geological evidence that Earth is far, far older. But for my opponent, the debate was not about the testable age of our planet. For him, it was about evolution. For him, it was about the apparently irreconcilable discrepancy between what our brains can observe along with the knowledge we can acquire and store, and the discovery that we got to this exalted state through the same process that makes some finch beaks short and sharp and others long and tapered. He just cannot believe it.

I sympathize with the troubling nature of the shortness of our lives, but a relatively short life is what we each have in store. Wishful thinking cannot change the facts, but scientific thinking can place them in a greater context. Human mortality can get you down and make you want to listen to old country western songs about how miserable life can be—or it can fill you with joy.

We have found out at least one, nearly incredible, truth about how we all got here. The astonishing thing about nature and the universe is that we can understand any of it. We humans have been around in our present form for nearly 100,000 years, yet almost everything we know about evolution has emerged in just the past 150 years. Think what lies ahead for our species, if we preserve biodiversity and raise the standard of living for everyone. We will make discoveries that would have astonished my grandmother, my colleague Bruce, and you and me today.

23

MICRO OR MACRO—IT'S ALL EVOLUTION

To understand evolution, we need to think both big and small. It's a recurring theme in evolutionary research, going back all the way to Darwin and Wallace. When I was a full-time engineer, I worked on drawing boards for huge airliners at one company and microscopic instruments at another. Down in the lower right of virtually all engineering drawings, there's a box where I, as the designer, indicated the scale of the picture. At Boeing, I drew at around 1 inch to represent 100 inches, 1:100. At Sundstrand, it was 1 inch to represent 0.010 inches, 100:1. It still fascinates me that the physics worked at every scale. A hydraulic actuator powerful enough to move a house is subject to the same laws of nature as a tiny spring that can detect the gravity of the Moon.

When it comes to evolution, we need to consider the big picture and the small picture. Nature affects every individual, but the effects of natural selection become apparent on the large scales: on groups, populations, species, and whole ecosystems. Researchers have coined the terms *microevolution* and *macroevolution* to describe the different ways in which evolution can unfold, though both are guided by the same fundamental principles that lead from micro to macro. Today, especially in the U.S., there are creationists who indoctrinate people to think only of the small picture. They accept the micro but reject the macro, because micro is all that their faith can accept. It's sad, and it's not science. The natural world is a package deal; you don't get to select which facts you like and which you don't. And in this case, you can't understand one kind of evolution without the other.

In its original form, Darwin's natural selection describes what happens to an organism once it is in the environment. For us, the process kicks in once we're born. For a plant, it would be once the seed is out there. For a fish, it would be once her (and his) eggs are deposited. Each new generation may be able to exploit its world's resources or not. It may get lucky and come upon plentiful resources, chemical energy for bacteria, a nice wetland for a frog. You are born with a set of genes copied from your ancestor or ancestors. Being able to keep warm, or cool, or digest the food resources around you, these are the forces of selection. There are other ways things can change from generation to generation, however.

Whether you get eaten or killed (get unselected) before you reproduce (get selected) is a huge driver of genetic change. But your genes can also be different from those of your parents just through random mutation, which is the imperfect copying from one strand of DNA to another. It can literally be a cosmic ray from outer space that knocks into one of your genes and changes it. Genes sometimes jump from one place on the DNA molecule to another. These are called transposon genes. It could be that your parent's eggs or sperm (or a plant's ova and pollen) got messed with a little by some chemical. It could be radiation from some radioactive elements in Earth's crust that caused a mutation. Sometimes viruses get into the reproductive cells of an organism and modify its genes. Virus manipulation can also be exploited deliberately—to adjust the genes of corn plants so they are tolerant of aggressive weed killer, for example.

With the little changes that happen with reproduction, the configuration of your DNA and your genes can change for you and others in the population of your kind. This is called "genetic drift." If the genes drift a little at the same time that there's a change in the environment, the drifted genes may be the only ones that make it through. This is an example of microevolution—a change in the genetic mix within a species or population. You can think of it simply as evolution by small changes over short times.

Another source of change is the random variation that is amplified in small populations. Consider a bag of Halloween candies, where half of the candies are orange and the other half are brown. Reach in the bag and take out a handful. In general, the smaller the sample you grab, the greater the chance that it won't be an even mix. If you grab five candies, you might have three orange and two brown. That would be a 20 percent discrepancy favoring orange. On the other hand, if you were able to grab five hundred, whatever unevenness was in the mix would be smoothed out. Your ratio of orange to black would be much closer. You would seldom have such a large 20 percent difference. In the case of genes in nature, if something happens in the environment that leaves you stuck with only a small handful of individuals, the equivalent of a small handful of candies, from that moment on your uneven mix will be favored. It will be the set of genes that gets carried on. Since in this case, we are only considering one species, evolutionary researchers often refer to this, too, as microevolution.

Along this line, let's say there's a favorable mutation in a population, such as a slightly darker skin pigment that provides better protection from ultraviolet light from the Sun (which happens now and again). People with that mutation might do well and reproduce more often in places with a lot of ultraviolet exposure. That favorable darker skin gene will preferentially show up in their kids and grandkids and great-grandkids. Because it's so important to human cultures, I've devoted all of chapter 32 to this phenomenon. Researchers call this "gene flow." The gene flows out into the population, albeit over successive generations. Since we're just talking about one gene in one organism, researchers often describe this as microevolution, as well.

In contrast, we have macroevolution: sexual, artificial, and natural selection writ large. Instead of considering one gene in one organism, we are now looking at sweeping species changes caused by shifts in the environment or by mass extinction events. The processes of microevolution and macroevolution are fundamentally the same, only the scale is different. We can study individual genes, a sequence of chemicals along an organism's DNA molecule. Or we can study a population of organisms, or an ecosystem full of organisms. They get passed on or eliminated by the same rule: If you fit in, whether as a gene or as a whole *Tyrannosaurus*, you'll get carried on. If you don't, you won't.

This distinction between micro and macro is useful for evolutionary biologists. Where it's been troublesome for me as a science educator is with the creationist community. As I learned firsthand from my debate with Ken Ham, an especially committed gentleman, creationists will seize on anything they see, or claim they see, as a loophole in evolutionary science. Mind you, finding flaws in a scientific theory is a noble business. It's a crucial part of the process by which science advances. But it requires honesty and consistency, and I don't see much of either in the creationist arguments.

Creationists will say that they accept microevolution—for instance, the way one strain of virus can mutate into another, a mutation for which our body's library of antibodies finds no match. It would be hard to deny that, since it happens every year at flu season. But then they'll turn around and insist that we have nothing in common with the other creatures here on Earth, because a higher power must have taken the time to make humans different. They use the term *microevolution* for the part of evolution that they find theologically acceptable. They use *macroevolution* for the part they don't like, perhaps because it's too disturbing. For them, macroevolution can't be real, because it cannot be reconciled with their belief that they are special, chosen for special treatment.

As you may know, creationists have gone to extraordinary lengths to reconcile their beliefs with the world as we see it. They've invented terms and written stories to try to make the incredible credible. Reading *The Bible* as written in English, they've coined a nominally Hebrew-derived word *baramin* to describe 7,000 kinds of plants and animals that were supposed to be on a boat 4,000 years ago and have since developed (through microevolution . . . that somehow turned macro) into 16 million species. They even use the phrase "mutation selection" to describe processes we all can observe in nature, like the London Underground mosquito speciation. But they don't accept the big picture of our common ancestry with every other living thing and the evidence for deep time and the age of Earth.

Here's the punch line: None of the micro ideas make any sense except in the context of the macro, and vice versa. Microevolution is just the raw material from which macroevolution occurs. This is a case in which scientific concepts become confusing, even misleading, when taken out of context. The bottom line is that nature doesn't care what words you use.

Creationism strikes me as an astonishing waste of time and energy. I would love to be able to ignore it and focus on the real science, but creationists work very hard to disrupt science education and force their weird worldview on our students. So let's make the best of an unfortunate situation, and use the creationist attacks as a learning opportunity. When you hear the terms *microevolution* and *macroevolution*, be attentive to who's using them. Think about how evolution works, on all scales of space and time. Viruses mutate from day to day. Fish evolved into land animals and eventually begat dinosaurs and blue whales, over hundreds of millions of years. It's a beautiful, complicated story on all scales.

So please: Think big, and think critically.

24

MICHAEL FARADAY AND THE JOY OF DISCOVERY

If you believe public opinion polls, about half of the American public does not accept the proposition that life on Earth—including humans—is the product of billions of years of natural evolution. At the same time, these same people seem to accept everything else that scientific discoveries and diligent engineering bring us. They don't doubt the chemical synthesis in their food, the electrical physics in their smartphones, or the relativistic corrections (Einstein's theory of relativity) that keep their GPS (Global Positioning System) signal accurate. Perhaps, as I speculated earlier, fear is part of what holds many people back from embracing evolution. If so, that puts a special responsibility on the scientists and those of us who write about them. If fear is pulling people one way, then we have a public responsibility to pull people back the other way and offer something just as powerful, something wondrous.

Too often, this is not what happens. I have met a great many people, who have told me that they were exposed to science in a joyless way. They were forced to learn about science as a series of obscure facts with a bunch of confusing equations to memorize. They were given a general sense that the world is difficult and a bit annoying to see through the eyes of a scientist. My, oh my, do I have a different view of the world. Let me tell you a story that still drives me wild.

From time to time, people ask me: If you could meet anyone in history, who would it be? I have a quick and easy answer: British genius Michael Faraday, the man who learned how to put electricity to practical work and one of the greatest science communicators in history. By 1800, many scientists across Europe were busily experimenting with electricity. Alessandro Volta built piles of plates separated by saltwater-soaked cloth or cardboard. The plates were alternating layers of copper and zinc. Such piles produce what he called electro-motive-force, which we now call voltage. André-Marie Ampère showed the relationship between the intensity of electricity and the magnetic force it produces. We now have the unit of electric current called the ampere, or amp. Michael Faraday figured out many of the crucial details about how electricity works. Among other things, he built the world's first electric motor. He also shared his ideas with the public in an exuberant style.

In 1825, Faraday began a series of lectures at the Royal Society in London called the Christmas Lectures. These were for a general audience, including kids, and have been delivered every year since, except while London was being bombed during World War II. Carl Sagan was one of the more notable modern Christmas Lecturers, in 1977. These are lectures with words, to be sure, but they are also performances featuring wonderful science demonstrations and showmanship. After several years of investigation in his laboratory, Faraday perfected a demonstration that goes like this: On a lab bench or table, about two meters (6 feet) long, Faraday set up two coils of wire that were connected by parallel wires, like a toy train track with tunnels at each end. In the center of one of the coils, on a suitably shaped block of wood, he set a magnetic needle compass.

Magnetism had been known for centuries; Christopher Columbus relied on a magnetized needle mounted on a suitably shaped floating block of cork to guide his ships. But Faraday took magnetism into new territory. With the compass in a coil at one end of the bench, Faraday moved a bar magnet in and out of the other coil at the other end of the bench. The compass needle moved. You can try this and get the same result. A magnet moving at one side of the stage caused a magnetic needle to move at the other end of the bench.

This may sound like no big deal. Many of you reading right now may have played with magnets and compasses (and perhaps ruined a few with too much magnetism too close to the delicate needle of the compass). It's reasonable that others had noticed that when electricity flows through a wire, a magnetic field is created around the wire, which can easily influence a compass needle. But Faraday realized what apparently no one before him realized: The process also works the other way around. If you have a magnet move near a wire, you get electricity in the wire. Faraday observed and carefully described the key idea. It's not that there's a magnet; it's that there is a *moving* magnet, a *moving* magnetic field.

At the Christmas Lecture, Faraday didn't just put a magnet near the wire. Instead, he made the magnet move, which in turn created a moving magnetic field. His audience was captivated, and so were his fellow scientists. Almost everything you touch and see all day long owes its existence to Faraday's discovery, because this is how we generate electricity. Just look around; what would be in your field of view without electricity? Hardly anything! Lights, televisions, computers, refrigerators, and coffeemakers would not otherwise be here. Everything that's manufactured—tables, chairs, cars, streets, carpets, tiles, clothes—now depends on electricity. Food is raised on farms that depend on machinery and transportation systems. This book was written using electricity and published using electricity, whether you are listening to my voice, or reading on paper or on a screen.

Saving the best for last: A woman came up to Michael Faraday after he performed this demonstration and asked: "But, Mr. Faraday, of what use is it?" Faraday famously replied, "Madam, of what use is a newborn babe?"

I hope you can feel Faraday's incredulity. You've got to admire him for not coming unglued. He could have said something like, "Lady, are you daft?! This is not a big deal to you!? Did you notice, for cryin' out loud, that I am not . . . I mean no one is . . . touching the compass?! Some force is passing from this end of the table to that end of the table, and that piece of metal is moving as though a witch were over there casting some sort of spell! By the stars, woman, this is utterly astonishing . . ." William Gladstone, when he was Chancellor of the Exchequer in England, reportedly asked Faraday a similar question. This time, Faraday responded—with perhaps more malice—"Why, sir, there is every probability that you will soon be able to tax it."

When I reflect on this story, I keep thinking about how much pleasure Faraday took in demonstrating his discovery and sharing it with the world. He got joy from it every single time. Darwin, I'm sure, felt the same joy of discovery. But it was different for Darwin than it was for Faraday, because Darwin's discovery was so troubling for so many on religious and philosophical grounds. The discovery of evolution led us to a line of thinking that, for many, diminishes our importance in the scheme of things. Darwin himself wrestled with the implications of his discoveries, as did his devout wife. Faraday had no such complicated emotions toward electricity. He also, like few scientists before or since, thrilled at sharing his ideas with those around him. His lectures were packed, exciting events because he spoke about science with unfiltered passion, and without a trace of pretense or jargon.

Evolutionary science could use a spokesperson like Michael Faraday. For me, learning about our place in the great chain of life is anything but sad. For me, and I'm sure Faraday would have agreed, scientific discoveries are joyous. In studying evolution, we find the hidden explanation for tyrannosaur fossils, our tailbones, and the common cold. Here we uncover the secrets to life on Earth. It is science, but it is a process driven by the human spirit. For me, there is nothing more exciting. It reminds me of another famous Faraday quotation: "Nothing is too wonderful to be true, if it be consistent with the laws of nature."

25

MEDICINE AND YOU—EVOLUTION AT THE DOCTOR'S OFFICE

My father and his high school buddy Phil were excellent Boy Scouts. They could both start a fire in the rain. They could tie knots that most of us have never heard of—blindfolded. When I was a little kid, Phil's wife got skin cancer; her face swelled up horribly. All his practical talents were useless to combat this disease. So were all the talents of medicine. Phil's wife was a serious Christian Scientist who believed that if someone got sick, the affliction could be overcome by seeking the aid of a divine power. She refused treatment, the cancer metastasized, and she died. It was hard on everyone; the experience broke Phil's heart, and my dad's and my mom's hearts, too. Even back then, Phil's wife could have had the malignancy removed and probably would have lived decades longer. Nowadays the gap between what doctors can do and what Phil's wife would embrace is even wider. Medical treatments have improved drastically, and evolutionary research is a major reason why.

This is an important aspect of evolution that many people don't appreciate. Evolutionary science is not just about the history of life.

It is a research program with very immediate, tangible benefits. It guides modern medicine. For instance, we've discovered that cancer evolves. Cancer cells can mutate in the body of a patient, so that malignant cells find new ways to get a supply of blood and become resistant to our anticancer drugs. We can use hormones from other animals—like insulin derived from pigs—to treat people because we came to understand our common ancestry. Medical researchers create new vaccines every summer to anticipate the evolved, mutated flu virus that will make the rounds in the autumn. The connections go on and on.

In one form or another, humans have been practicing medicine for millennia. African tribes drilled holes in each other's skulls to relieve fluid pressure. First Nation tribes in North America developed several pain-relief medicines. Humans everywhere developed techniques for the treatment and healing of broken bones. It is a consequence of having enough brainpower to figure out causes and effects in our own human bodies. The medicine we practice today is fundamentally different from the way that humans dealt with diseases and injuries through most of our history on Earth, however.

The difference is that today's medical practitioners can draw on predictions made by our understanding of evolution. As any biologist will tell you, living things everywhere on our planet have an astonishing number of things in common. We are made of cells. We have the instructions to build any one of us in almost every one of our cells; we all have DNA. We all reproduce in the midst of all the world's other living things. As we do, subtle changes get built into each succeeding generation. Today's drugs and vaccines are possible because of these insights.

Recently, the evolutionary aspects of medicine have taken a surprising turn. Doctors are starting to look at evolution not only as something that affects human health from the outside, but from the inside as well. They are thinking of each person as a walking, evolving ecosystem. I admit you might not think of yourself as an ecosystem. At least, not yet.

Most of the living things on this planet are made of a single cell with no nucleus in the middle. Most Earthlings are microorganisms. You probably have no trouble accepting that. Now try this: Even the majority of the cells *in your body* are microorganisms. They outnumber the cells of your body by 10 to 1. Those microbes are living, metabolizing chemicals, producing waste chemicals, and interacting with each other. Collectively, they are known as your microbiome. You are their ecosystem. It's wild.

When we're born, none of these organisms are on board. Babies have no microbiome; there's no system of complex biological activity in their digestive tract. They get their microbiomes from their parents. All the snuggling, and kissy facing, and breast-feeding enables countless bacteria to make their way into an infant's tummy, where they live and reproduce for that new human's entire life. The ecosystem or microbiome in our digestive tract coexists with us. We depend on it. Much of the food we eat is broken down by these bacteria. When things go wrong with our intestinal microbiome, trouble starts.

I have a feeling (and it's not a good one) that everyone reading this book has been horribly sick at some point in his or her life. Many species of bacteria produce waste products that are toxic. Our bodies have systems set up to detect and expel the sickening toxins. We vomit—we expel—which usually, but by no means always, does the trick. Bacteria are here today because they were able to spread themselves around. The expectorated fluids from inside the body help spread bacterial material. This seems obvious now, but it is a fairly recent discovery. Even 150 years ago, back in Darwin's time, people weren't yet sure that microorganisms could make you sick or kill you. Biologists discovered the microbiome just over the past couple of decades.

They are still puzzling out how the microbes in your body help keep you healthy and well nourished. A person's microbiome might even be an important factor in controlling his or her obesity.

The discoveries of immune responses and germs have changed the world. The discovery of the process by which all these germs have come to be has enabled us to embrace the good and fight the bad. Consider the following: We have developed dozens of antibacterial drugs. These are molecules that break down or pierce the cell walls and membranes of bacterial pathogens. But because bacteria are continually reproducing, they are also continually mutating and evolving new defenses. Through the filter of survival of the best suited, many of the bacteria in our environment today cannot be stopped by antibiotics that worked effectively just a few years ago. If we want to continue living healthy lives, we are going to have to come up with new ways to fight bacteria. That will take a deep understanding of what goes on in the bacterial world. You will not be shocked to learn that it has to do with evolution. More about that in the next chapter.

There are an absolutely astonishing number of bacteria on our planet. By most reasonable estimates, there are about a million-trillion-trillion, or 10^{30}, of them (ten followed by 30 zeroes) here on this planet. There are more bacteria on Earth than there are stars in the observable universe. Like every other living thing, bacteria exploit the same chemicals for nourishment. And so, they compete. They compete like crazy. To that end, they fight each other. They fight not with spears and stones, but with poisons. Bacteria produce poisons to kill or severely inhibit other bacteria. By long word-coinage tradition, these toxins are called "bacteriocins." They are able to "cut" bacteria. There's a bacteriocin specific to *E. coli*, which goes by "colicin." Get it? Cut coli? Making up words like that is a popular business in biology.

In general, the known antibiotics are toxins or chemical inhibitors produced by other types of organisms. Penicillin, for example, comes from a fungus, which is quite a bit different from a bacterium. Somewhere along the way, the mold *Penicillium notatum* chanced upon a combination of chemicals that disrupts the cell walls of a great many bacteria, which enables the fungus to advance its hyphae (fungal tendrils) without getting attacked, at least not successfully attacked. Being an organism that is different from a bacterium, a fungus can carry an antibiotic—in this case cell-wall-disrupting chemicals. But if you are a single-celled bacterium producing a chemical within your cell membrane and wall that would split the membrane and wall wide open of another closely related bacterium, it gets complicated.

With billions upon billions of reproductions, bacteria have come upon chemicals, bacteriocins, that attack other bacteria with proteins that are specifically shaped or keyed to disrupt the cell wall of just one or a very few other types of bacteria. These proteins don't attack the cell wall of the bacterium producing it . . . That wouldn't work. A protein is a workhorse molecule produced by living things; the chemical properties of a protein derive in part from the molecule's shape. It's not just that a protein might carry a nitrogen atom that might bond with an oxygen atom. It's that the protein holds or presents those atoms in such a way that they react only with other molecules that can or do receive or interact with them; these proteins fit like keys in locks.

Now when I say that bacteria couldn't do this one thing so they do that other thing, I should clarify. It's not a choice as such. Bacteria produce proteins, all kinds of proteins. These are the molecules that give things like your bones, skin, and hair their shape and structure. If a bacterium happened to produce a protein that split its own wall open, well, that organism would die. But by having billions upon billions of proteins reproducing for millions upon millions of years, proteins get produced that serve as bacteriocins; they destroy or kill other bacteria, albeit only bacteria of a very specific type.

With the discovery of this remarkable property of many bacteria, scientists have sought to produce antibiotic-style bacteriocin drugs that attack just one specific type of bacterium. Suppose you got sick with some miserable staphylococcus infection, and the particular strain of staph you got has been around a long time. Its bacterial ancestors have encountered the human-made antibiotic drugs for decades, and the descendant strain that your body is fighting is resistant, or largely unaffected by the cell-wall-disruptive qualities of our penicillin, or erythromicin, or what have you.

You might start to lose. The bacterium might start to produce so much toxin that you can't fight it. But then scientists take samples of the particular strain of staph from your mouth, for example. Next they breed a particular type of bacterium that happens to produce a particular type of bacteriocin that happens to kill the staph bacterium that's infecting you. An agency like the Centers for Disease Control and Prevention (the CDC) could in turn breed a group of closely related bacteriocin-producing bacteria, or administer the isolated bacteriocin protein on its own, which you could drink like a glass of orange juice. The bacteriocins produced inside you either by the special strain of anti-staphylococcus bacteria or by the bacteriocin itself, and you'd recover in no time.

This line of evolutionary attack sounds seductive, but it works only if we can identify the specific bacterium and its cell-wall-disrupting bacteriocin. Researchers in southern Russia have been doing it for years. Along with attacking infectious diseases with bacteriocins, their research has involved attacking skin infections. By identifying just what bacterium is infecting a patient, these researchers have managed to come up with the right bacteriocin-producing bacteria that can produce enough of the bacteriocin protein to destroy the infecting strain.

A future that makes this technology available to all of us would be a bright one indeed. It would be a life-affirming result of our understanding of evolution, and the many historical steps that got us to this point. First scientists including Anton van Leeuwenhoek, the seventeenth-century Dutch microscopist, discovered the microscopic world. Then other scientists discovered that our immune system could be trained or induced to fight specific diseases that it's exposed to. Then scientists identified specific bacteria and specific viruses. Then scientists discovered chemicals or molecules that disrupt cell walls and membranes of certain bacteria. Then scientists discovered that bacteria fight each other. For example, way down in your intestines, there are generally three different types of *Escherichia coli* fighting each other with specialized bacteriocins all the time. This is what researchers mean when they talk about "standing on the shoulders of giants." Science is a beautifully cumulative process.

For about two hundred years, humans have used other animals to test the effects of medicines and medical procedures. You may have heard of the Rh factor in blood. The term comes from rhesus monkeys, in whom it was discovered and studied. You may have been troubled or grateful for procedures in which eye makeup is tested on rabbits before its release to market. You may know about mice that are given certain doses of certain food additives or cigarette smoke to test for ill effects. You've probably used the term *guinea pig* to describe someone who goes first in a procedure in which the outcome is unknown.

All of these test animals can be used to see what would happen to us if were placed in the same situation for a simple reason: At the cellular level, humans and monkeys and pigs and mice are very much alike in construction or design. We share almost all of our biochemistry. We all have DNA, and it's nearly the same. In rhesus monkeys, we're close to 93 percent the same. In mice, it's closer to 90 percent overall. Just think about the potential consequences of these numbers. If you're a germ, you might be able to move from species to species. That's why we worry about, for instance, avian flu and swine flu. Or alternatively, 10 percent may be enough to invalidate any conclusions about infections that you might draw. In either case, human researchers can use our understanding of mutations and natural selection to determine how much of what we observe in these animal models applies to us.

Every one of us in the developed world has benefited directly from the medical tests performed with these animals. It's yet one more scientific achievement based on evolution.

In many ways, we have barely begun learning how to fight back against disease. Just think what else we don't know about bacteria and their interactions. With each step in the process, we used the method of science: Observe. Hypothesize. Predict. Experiment. Compare what you expected with what really happened. This rigorous form of the scientific method—one in which very disciplined experiments are conducted under carefully controlled conditions—really is a result of medicine. In order to observe and isolate the effects of germs, for example, you have to look very carefully, because you just can't see them without isolating them carefully and peering diligently through a microscope. Evolutionary theory benefited from medicine, and now medicine benefits from evolutionary theory.

As I look back on this particular human endeavor, I'm astonished at how careful these scientists were to isolate smallpox, rabies, mumps, measles, whooping cough, and rubella. We might say that we are a lucky bunch to be alive right now. But it wasn't luck. It was a strategy; it was science. You are here because a large number of people worked together around the world and across the centuries to understand how the natural world really works.

26

ANTIBIOTIC DRUG RESISTANCE—EVOLUTION STRIKES BACK

Do you get a flu shot every year? You should, because influenza (flu) viruses are evolving right under your vulnerable-to-infection nose, and you need to keep up. Viruses make their living by infecting living cells and inducing them to make copies of the virus. Those replicas pour out and encounter nearby cells, which results in more infection and yet more copies of the virus. If you think that sounds like war, you're not far off. In the winter of 1918–1919, the Spanish Flu killed about 50 million people—more than all of the combat in World War I, which had only just ended. Keep in mind that there is no evidence that viruses are malicious. They are just following their evolutionary path, multiplying wherever they encounter a survival advantage. The frightening thing is that the mindless, relentless drive of natural selection is now overwhelming our best defenses, making once-tamed diseases dangerous all over again.

Humans come to the battle equipped with their own defense, also shaped by evolution: the immune system. It's a complicated set of chemicals and processes that your body carries with it to fight diseases brought on by viruses, germs, and multicellular parasites. It learns from experience, developing targeted defenses against every infection as it occurs. So, let me pose this question: If our immune systems were working at a normal pace or performance level, wouldn't we have overcome every infectious entity our bodies have ever come across, or have come across us? Wouldn't we have defeated every germ and parasite in nature? As I'm sure you're aware, we have not. Nature doesn't work that way. It's one of the least pleasant consequences of evolution.

Over the course of your life, you've undoubtedly been infected by colds, flus, stomach viruses, food poisoning bacteria, and nature only knows what all else. Yet, neither you nor anyone else has overcome all these threats. You may even be in the good habit of washing your hands frequently to keep from getting sick again. You know intuitively that there are more germs out there that your immune system has never seen, and there will be the rest of your life.

So, let's pose the next logical question. Where do all these new never-before-encountered germs come from? There is no reason to look for a biowarfare factory in some undisclosed location that produces new germ designs and unleashes them on the world. Instead there are as many germ factories as there are people. Each of us serves as an incubator for new strains or varieties of germs.

Because of their complete lack of nerves or brains, germs are not mean-spirited or ornery. They just are what they are. Since the beginning of bacteria here on Earth, at least 3.5 billion years ago, there must have been configurations of molecules that caused those bacteria trouble. Like every living thing we have ever found on Earth, viruses have long-stranded molecules that carry the genetic information a virus needs to force or induce bacterial cells to produce more of these same viruses. Since they're made of the same molecular stuff, and they are so very much alike on the molecular level, it is reasonable to infer that bacteria and their viral enemies, the bacteriophages (the viruses that attack bacteria), came into existence about the same time. Hence my inclusion of the Vira as a domain of life. Each organism has vied for the same chemical resources since Earth cooled off enough to have liquid water extant on its surface.

Among the fascinating aspects of bacteriophages is how specific they are. Only certain phages attack only certain bacteria. As I mentioned early on, a reasonable explanation for this is that they all came into existence about the same time. The key would be, as the car dealer often mentions on a television ad, volume, volume, volume. By producing enormous quantities of bacteria and enormous quantities of phages, the chances that they will find each other gets high enough to sustain entire ecosystems. That's part of what is going on within your microbiome.

Let's peer back into history to see how we got here. Bacteria in the primordial seas were undoubtedly attacked all the time by phage viruses, etc. There would be billions and billions of each. The key is that, when copies of molecules are made, mistakes or imperfections result. When a virus has managed to infect thousands of cells, each of which is producing thousands of viruses, that virus's ribonucleic acid (RNA, a relative of DNA) is going to get replicated with some mutations. Expand that into the future where we have tens of thousands of humans getting the infection, each of us in turn producing millions of viruses. Sooner or later a mutation is going to emerge that can infect, or reinfect, most of us.

Do we have to accept, then, that we are going to continually be infected by strange new strains of viruses, bacteria, and parasites our entire lives, and there's nothing to be done about it? Well, in the same way we take steps to avoid getting injured by trees we're cutting down, or traffic we're crossing, we take steps to avoid getting infections. We wash our hands and avoid sick people, at least to the extent possible.

I remember well being kept inside during certain summers to avoid being exposed to the virus that causes debilitating polio. So it is logical that we, as a scientifically literate society, take steps to create immune responses inside our bodies as well as outside. We can, because we understand evolution.

With all the mutations going on, crafting a flu vaccine is like shooting a needle at a moving target. Every winter, when flu season begins, the viruses that are circulating are slightly different than those from the previous season. That is why the CDC works with the Food and Drug Administration (FDA) and the World Health Organization (WHO) to anticipate the next season's troublesome strain of flu virus. The general trick is to capture or get samples of the flu viruses infecting people in the Southern Hemisphere during their winter and prepare vaccines. It is an evolution-driven health system that almost everyone takes for granted.

Pathologists—scientists who work with and study infections like the flu—prepare a vaccine either by using a virus that has been weakened (attenuated) a little bit with a chemical reagent, or by using a dead or completely disabled virus that is still structurally intact. It still has the right configuration of proteins for your body to recognize, so it triggers an immune response, but it is not capable of causing an infection on its own.

The first order of business for the immune systems is to recognize incoming viruses, bacteria, or parasites that are unwelcome inside us, because those pathogens have schemes for taking control or hijacking our cells to do the work of making more viruses or bacteria. (Along the way, the immune system needs to ignore benign microbes, as well as the body's own cells.) Once your body recognizes an unwanted intruder—an infectious agent—it can send antibody proteins to wrap up the virus and pry it apart. To do this antibodies have to be attuned to a pattern of proteins on the outside of the invader. If your immune system, with its antibody molecules, does not recognize an infectious pathogen, your immune system does not respond, at least not right away. That gives the virus, bacterium, or multicellular parasite a tremendous head start, and can make you very sick.

The reason these infectious organisms are still here is that they're always changing, always mutating as they reproduce, taking on new chemical identities that the immune system does not recognize. They are always evolving in a way absolutely consistent with what we predict per evolutionary theory. Germs and parasites evolve with remarkable speed in geologic terms or in comparison with the amount of time life has been on Earth. That's their business. We humans, as their slowly evolving victims, have to make it our business.

Bacteria take chemicals out of the environment to run their metabolism. After consuming the right amount of the right chemicals, bacteria have enough chemical energy to reproduce. They do that by splitting themselves in half. Of course, it's a bit complicated biochemically speaking, but the idea is just that. We call it binary fission—splitting in two. That would be fine for the bacteria, but as they reproduce and metabolize the chemical environment around them, certain of them produce miserable toxins that make us sick.

In some primordial age, a bacterium chanced upon a way to produce toxins that made one of our ancestors sick. That was a great day for bacteria-kind, because they had come across an effective way to spread themselves around. Our ancestor may have had so much toxin in her nose that she sneezed on others in her tribe, spewing live bacteria with her saliva and mucus. Or she may have been just breathing air laced with bacteria-bearing water drops in the stream. Or, *ugh*, bacterial toxins may have produced diarrhea. And as our ancestor's body was, uh, ridding itself of toxin, it was also spreading the bacteria. Once a bacterium or family of bacteria came across this scheme, it's hard to stop. But human scientists came across a few very effective techniques.

Bacteria have a cell wall that keeps the microbe separate from its environment. Within the wall is a membrane that keeps different organized subsystems or "organelles" within the bacteria separate. Humans have found a way to make the proteins, the molecular structures that hold the bacterial cell wall together, fall apart or pry open. When that happens, a bacterium spills its guts; it falls apart and stops metabolizing and producing toxins.

The molecules that pathologists craft and chemical engineers work to mass-produce in oil-drum-sized "reactors" are what we generally call antibiotics. Perhaps the most famous among them is penicillin. You may have heard the story of Alexander Fleming, the Scottish biologist who noticed in the 1920s that the *Penicillium notatum* mold can kill the *Staphylococcus aurius* bacterium that causes staph infections. You've definitely heard of the bacteria-slaying compound he isolated from those molds: penicillin. Fleming realized that if this mold could be isolated and produced in large quantities, it would be very effective against bacterial infections. Penicillin ended a great deal of misery, and it saved countless lives.

Along with those measurable benefits, penicillin led to the development of dozens of other fantastic antibiotic drugs: ciprofloxicin, polymixin, tigecyclin, and a great many more. Some of these drugs keep bacteria from dividing or reproducing; others kill the bacteria outright. Either way, antibiotics have changed the world and completely changed the expectations of modern medicine. Our most dangerous enemies, germs and parasites, were all at once easy to kill and overcome.

But the process of evolution has rendered a great many of these drugs far less effective than they once were, even useless. Since bacteria are able to reproduce so quickly compared to the organisms they attack and exploit, they also mutate quickly. Over just a few years, genes that conferred resistance to the drugs were randomly created or acquired. The result: A tremendous number of bacteria that were once quite controllable are now troublesome at best, deadly at worst. Evolution is enabling them to bite back, and that's bad for us.

Medical professionals nowadays have come to realize the gravity of the problem, and have been warning the public that overuse of antibiotics is making them ineffective. The more of these drugs we use, the more chances infectious bacteria have to come in contact with them, and the more likely these bacteria are to come up with descendant strains that are able to chemically protect themselves from the effects of the drugs.

I have participated in campaigns encouraging people, especially anxious parents, to not overuse these wonderful products of science. You may be among those who have, or who know someone who has, had a sick kid. The parent takes the kid to the doc and is so aggressive or concerned or sympathetic that the doctor prescribes antibiotics for a disease that the patient's immune system would have beaten eventually anyway.

Furthermore, because of their completely different chemical mechanisms, viruses are unaffected by the drugs that target bacteria. It has to be a bacterium that's infecting you for an antibiotic to do any good. (Don't ask your doctor for an antibiotic if you have the flu; it won't do a thing except help breed more dangerous bacteria.) Viruses have no cell walls to breach. They have to be attacked with antibodies that are tuned to find them, and find them fast. That's why the medical community has sponsored campaigns encouraging consumers and health care providers to "Take the Right Drug for the Right Bug."

It's also important to finish your antibiotic prescription. That is to say, if your doctor gives you a two-week course of antibiotics, be sure to follow it all the way through, even if you start to feel fine in just a few days. Otherwise, you will probably not have killed all the bacteria that are in you, and the ones that are still hanging around will probably have a tendency to be immune to the antibiotic. They may not have complete immunity, but a fractional resistance will be passed on to subsequent generations of your germs, and they will be resistant to the same drug after you've passed your infection along to some other hapless victim and soon-to-be patient. You'll be aiding and abetting the enemy.

This all happens by the exact same evolutionary process that brought about all the creatures and ecosystems that are extant in the world today. It is modification by natural selection, it just happens to be taking place in bacteria. It is one more vital reason to promote universal science literacy: Evolution is a matter of life and death.

27

THE IRRESISTIBLE URGE OF ALTRUISM

I'll admit that the last couple of chapters have been pretty downbeat, and make evolution sound as if it's out to get us. Near as anyone can tell, evolution is not guided by a mind or a plan. It just *is*. Our perceptions of whether evolution is being generous or malevolent are based entirely on whether we think that we are the ones coming out on top—which, come to think of it, is a very evolutionary way to look at the world. So take what I'm saying as an entirely subjective message, but I think of this as a much happier chapter. It is about a topic that benefits all of us: altruism, the instinct that makes us look out for one another.

Let's start with an example from my childhood. Like my father, I was a pretty good Boy Scout. I was, and am, pretty good in the woods. I can split wood, start a fire, rustle up some grub, lash spars, set up shelter, and find my way out of the woods through unfamiliar territory. Along with the exploration activities, though, part of the deal with the Boy Scouts is to help out in the community. You're required to do a good turn daily, to help someone every day, even if only for a few seconds. One of my modern mottoes is posted on billnye.com: "To leave the world better than you found it, sometimes you have to pick up other people's trash." I guess my altruistic impulses came from my mom and dad. They did their best to leave the world better than they found it (you may have to forgive them both for begetting me, I suppose. My mom always said dad was a great dancer; I guess one thing led to another).

When I first entered the workforce, I volunteered at the Pacific Science Center in Seattle. I helped move boxes around on weekends, and I served as a "science explainer" now and then. It just makes a feller feel good to help out. The demonstrations skills I learned there led me to a whole career and the writing of this book. I was also a tutor through the "I Have a Dream" program. Both of these early jobs are examples of altruism, expending energy for the good of another. I like to think I was good at my work, but I always felt I got more out of it than the guests at the Science Center, or the students on Saturday mornings. I've had that experience many other times—when teaching a class or performing a particularly good science demonstration. I claim that I can sense it inside myself and inside others—that this type of altruism is encoded within us somewhere.

Religions often preach about the importance of doing things for others. There are countless not-for-profit organizations that promote the idea of service for the betterment of our neighbors, or for those less fortunate than we are. If you volunteer or serve others, I presume it makes you feel pretty good. Most of us take satisfaction in helping others. Most of us feel good when we believe that we've done the right thing. Just like you and me, evolutionary biologists call the phenomenon of doing the right thing "altruism," although they define it more rigorously. The word derives from the Latin words for "to this other." The origin and nature of altruism is one of the hottest areas of research in evolutionary science today.

In conversational language, altruism refers to being selfless, to helping another without expecting any reward. In evolutionary biology that idea is expressed in an equation: An individual provides or gives a service while gaining little or no benefit to her or himself. Specifically, the cost is greater than the reward or benefit. That relationship can be written very simply as

$$b < c$$

(benefit for the giver is LESS THAN cost to the giver)

There is always some cost in offering help, whether it is measured in time, energy, or risk of getting attacked. Meanwhile the recipient receives benefit—something like food, sharing the carrying of a heavy load, or a warning of an imminent attack by a predator—while the giver does not, at least not in an immediately obvious way. Scientists interested in the origin of this tendency in many animals, including us, have studied altruism in all kinds of intriguing ways. Much of that research circles around the basic question of altruism: In the battle for survival, why would any individual favor costs over benefits?

It's right about now that religions typically assert that belief in a higher power or faith is the root of altruism—that without a certain set of beliefs or inculcation into a religious doctrine, there would be no good deeds done in the world. At least, that is my interpretation of what I perceive as the positions of many organized Christian religions. But it is my view, and the view of many evolutionary scientists, that altruism occurs widely in nature. It shows up in animals like vampire bats and termites, for but two examples, whether or not they are exposed to religion.

Stick with vampire bats for a moment. They fly out at night and hunt for cattle. They bite the cow or bull's neck and gulp down blood that is loaded with nutrients for a mammal. They fly back to their roost, and through a series of signals, determine whether some individuals among them were unsuccessful that night: They flew and flew, but could find no neck to suck. In this case, the bats who did find food regurgitate some of the blood they gulped, allowing the hungry bats to get a meal. This might all sound vampirically creepy, but it is well documented and the way of a vampire bat's world. But the really strange part is the altruism part. Why would these bats do that?

As a survival-of-the-fittest first cut, you might see yourself in the role of a well-fed bat and remark (or think), "Why should I help this other bat, who flew so poorly and listened to that rock and roll music instead of his or her own ultrasonic echoes? He didn't eat? Well, to the dingy dark cave with him!" In other bat words, "Why not let him or her starve tonight? It would teach that bat a lesson about flying and echo-locating." As valuable a lesson as that might be, it's not how bats roll . . . or roost. When a fellow bat doesn't get enough food to drink, another bat helps out. Near as any of us can tell, these altruistic bats never went to Saturday or Sunday school. It's in their blood, pun intended. And studies show that impulse is widespread.

Let's postulate that altruistic behavior is ingrained in all mammals. Let's say, further, that all of us mammals have enough in common ancestrally to explain this. Somewhere a long time in the past, perhaps around the demise of the ancient dinosaurs 66 million years ago, bats and humans had a common ancestor and that group (colony or tribe) had a tendency to help a fellow out. Well, is that where it ends? If we found that ancestor, could we do some analysis of the evolutionary origin of altruism?

Probably not, because altruism most likely goes back a lot further than that. I've been to a remarkable place called Dinosaur National Monument. It's on the border between Utah and Colorado. The official location is Dinosaur, Colorado. One of their sayings is: "You'll want to stay here forever . . . the dinosaurs did!" How charming is that?

Apparently, about 150 million years ago, a river flowed through the location, and during a very large rain event there was an enormous flood. Hundreds of animals were washed downstream the way our human-built houses get washed away in floods. The animals drowned or were crushed in what might be called a bone-jam, like a logjam only more deadly.

If you've never been to the site, you should visit and see it for yourself. It's amazing. President Woodrow Wilson saw to it that the area was set aside as a national monument in 1915. Today, there is an enormous roof covering the main bone-jam area, known as the Carnegie Dinosaur Quarry. But if you walk around and you're alert, you'll find a great many other ancient dinosaur fossils and get a sense of the enormous scale of the deluge. As you consider all of the animals that must have been living in the same area upstream, you cannot help but wonder how they lived. Were they savages with no ties to families or homelands? Or were they like us, with tribal ties and their own kind of altruistic impulses? Did they have ancient dinosaur rituals like parades, birthday parties, and sunrise services?

I've also traveled in Montana and visited a couple of ancient dinosaur fossil sites there. It takes diligence, but scientists have found a great many dinosaur egg nests. Some of the most spectacular examples were buried by ash from what we now call a supervolcano under Yellowstone National Park. The nests here belonged to a duck-billed dinosaur called *Maiasaura*. You don't have to know much about ancient dinosaurs to see that they lived like birds in a rookery. The arrangement of the eggs in *Maiasaura* nests indicate that these animals cared for their young at least as attentively as a modern bird does. Observing modern birds, we can see that they have altruistic tendencies. Some species, famously the cuckoo, lay eggs in the nests of other birds, who tend to the alien hatchlings as though they are their own, apparently *just because*, just because it's the right thing to do. It's an amazing strategy that exploits other birds' altruism or just their bird brains.

The connection between dinosaurs and birds runs deep. Strictly speaking, all modern birds are classified as dinosaurs: Specifically, Reptilia gave rise to Dinosauria, which includes Sauropodia, which includes Avialae, the modern birds. Paradoxically, our birds are descended from the dinosaurs that are classified as "lizard-hipped" rather than those classified as "bird-hipped." It's yet another quirk in the history of scientific terminology. Modern birds have maternal instincts, and the ancient dinosaurs—nonavian dinosaurs, in modern evolutionary lingo—probably did, too. (A side note: *nonavian* is a poor term for those once-mighty beasts. Would you expect a sales guy to try to sell you a non-car truck? Sigh . . .)

Maternal care of offspring is not quite the same thing as altruism, but how many of us have had our best friend's mother treat us very well, just because we were kids who needed a snack or a ride or a few hours at the bowling alley? Does a human mother do that out of her maternal instincts or out of a humankind-wide altruism? If we allow the definition of altruism to include maternal care, then it becomes an even broader and older evolutionary impulse. A great many species that seem to be less thoughtful than we exhibit paternal care, too. I've been scuba diving around the Garibaldi damselfish. When you approach a nest of eggs, the male fish will go out of its way to annoy you until you move away. I've read accounts of divers being bitten by cute, happy little Garibaldis. Based on this experience alone, I'd say fish at least have a paternal instinct.

There is a good deal of evidence that certain fish also help each other out, albeit only in certain restricted ways. Consider the big-eyed squirrelfish and the reef wrasse. (Before we go on, I have always been charmed by our human-given whimsical fish names. A squirrelfish? Really? And, a wrasse would be an "old woman fish." That's just weird.)

At any rate, the squirrelfish swims into an area where the wrasse is working. The big fish lets the little fish eat tiny parasites off the big fish's scales. The squirrelfish even lets the little wrasse into its mouth to clean out parasites. The squirrelfish could just eat the wrasse, but it or they or he or she doesn't. The two species have a symbiotic or mutually beneficial arrangement. But, wait . . . wait . . . there's more.

Apparently, these two species of fish recognize each other as individuals. A certain squirrelfish seeks out a certain wrasse; they know each other in a fishy kinda' way. They help each other by not seeking other individual cleaner wrasses or another big squirrelfish. They guarantee each other's employment. But the altruistic part is that the squirrelfish chases away predators of the wrasse. The squirrelfish could just let the wrasse die at the mouth of the big bad predator fish, but instead puts its own life in danger on the little wrasse's behalf. It's an intriguing arrangement.

Altruism has a dark side, though. It requires enforcement at times, and that can involve punishment. Have you ever been mean, just because it somehow made you feel better? Have you ever sought revenge? The country singer Carrie Underwood had a huge hit with "Before He Cheats," in which the protagonist celebrates vandalizing her ex-lover's four-wheel-drive truck. She describes smashing the headlights, carving into the seats, and slashing the tires. The crowd goes wild when she sings it; I've seen it.

Have you noticed that it might take a lot of time, effort, and energy to exact revenge? Someone plays a practical joke on you, like faking a phone call from a big television production company telling you that the company does not want to produce a show called *Bill Nye the Science Guy*, which broke your heart and sent you into a funk for the rest of the week? (That example is fictitious of course and I'm using it just as an example!) Then after you found out it was just a radio DJ you'd met, who was simply pretending to be a television executive, you thought about all the different things you might do to make that guy's life miserable? (Not that any of this has anything to do with me, of course.) Even though it would take a lot of your time to get the parking tickets forged and set up the fake phone number for the fake law firm that would send him a bill for thousands of dollars? If you did all that, you are helping to keep our society in balance, and it is predictable using mathematical models of evolution.

The idea is that it costs you something to seek revenge—it takes your time and it's a little dangerous to smash headlights and carve seats. It takes time and money to produce fake tickets and set up a fake phone number or send phony invoices—you do it anyway, because making that other person feel bad makes you feel good. This has come to be called altruistic punishment. It is the flip side of evolutionary altruism. Evolutionary scientists claim that we are compelled to get back at someone who has wronged us because it's good for society. Our ancestors, who were not programmed to seek revenge (to gouge the paint job on someone's truck after she caught you on a date with a rival), did not pass on their genes. Revenge has an evolutionary purpose: It keeps everyone honest, so we can't resist seeking it. It's in the tribe's best evolutionary interest that you exact revenge on those who've broken the generally accepted rules. It's altruistic to punish.

Researchers have run many tests on the value and effectiveness of altruistic punishment. Classic experiments allow players to allocate money or points to other players in a group game. People never keep all the resources to themselves, apparently out of fear that sooner or later the other players will demand restitution if they do. It works every time. We are motivated to exact revenge or payment from someone we perceive to be a wrongdoer, even if it costs us money, time, or energy to do it—even when there is no apparent benefit to ourselves.

This is another case in which we can demonstrate that evolution influences not only the way we look, it also affects what we feel. The happy side of altruism helps the members of our human tribe directly. The mean side, altruistic punishment, keeps us all in check. We help each other out when we can, which generally makes us feel good. We act mean when we need to, which also makes us feel good—or at least better—and which, more important, helps keep the whole system of altruism running smoothly.

The big limitation of these experiments is that they are just tests. They are often simplified simulations of the real world. Influential nuances may be lost. And the limitation with underwater experiments is that these are just fish. It's hard to figure out exactly what motivates them by asking questions. Although they're always wagging their jaws, they refuse to talk, and their handwriting is terrible.

So biologists turn to playing games: More precisely, they turn to game theory, a system of thought that breaks down behavioral costs and benefits in specific scenarios. These games turn the idea of evolutionary feeling into something concrete and quantifiable. That's where we are going next.

28

GAMES SPECIES PLAY

My older brother Darby is great at card games. He usually beats me. In fact, he beats a lot of people. He has what is traditionally called "card sense." Without counting each card, he has an intuitive feel for how many face cards have been played, how many aces remain in the deck, and the likelihood that the certain cards he needs are in my hand and any other players' hands. Maybe my brother should become an evolutionary biologist, because theories of games have become very important to understanding natural selection, especially to understanding the altruistic impulses we just investigated.

The most famous of the evolutionary games is the Prisoner's Dilemma, formalized in 1950 by mathematician Albert Tucker, who specialized in what's called game theory. The game goes like this: Let's say two partners in crime are apprehended. Each is interrogated by the authorities; within the constraints of an imaginary situation, each can either tell the cops that the other guy (the accomplice) committed the crime, or each can admit that he committed the crime. The possible consequences are:

1) The first bad guy denies having anything to do with it and gets away free, so the second bad guy ends up taking all the blame.
2) They both deny involvement and both go to jail.
3) They both admit doing it, and both get off with half the penalty.

Test subjects have been put in a variety of situations simulating this situation. Computer programs have been created to simulate this situation. Mathematicians who specialize in game theory have analyzed this scenario as it carries forward. Each prisoner has to make the choice between denying and admitting. The game's outcome is resolved once the second prisoner makes a choice: admitting after the first has denied, denying after the first has admitted, etc. In many versions of the game, prisoners make their choices before they know what choices their accomplices have made.

In the simplified context of the Prisoner's Dilemma, it would seem that the best course for either prisoner would be to simply deny, deny, deny. In reality, the trend goes somewhat the other way. If the other guy denies, the first guy admits. If the other guy admits, the first guy also admits. Why would he admit if the other guy has already done so? It's not what you might expect of a hypothetical prisoner, who was only interested in him or herself. It turns out that humans are somehow biased to cooperate. They are prone to be somewhat altruistic, somewhat like the squirrelfish and the wrasses I was talking about in the previous chapter. Game theory is just another way of probing the ways in which evolution has instilled altruistic instincts in all of us.

If you want to really get inside those instincts, try examining your own responses to a few game theory–type thought experiments that have more of a real-world twist. The following scenario is rather different—it involves danger and rescue—but like the Prisoner's

Dilemma it shows how strongly we make choices that are driven by our inherited evolutionary tendency to be altruistic.

Suppose your house or apartment is burning. You are in the house with your kid and your mother. In this admittedly creepy scenario, you only get to save one. You are only allowed to carry out either your kid or your mother. (The cat and dog are on their own.) Everyone picks his or her own child. Now, suppose it's your mother and your brother's child. Everyone picks the kid. Now suppose it's your mother and a *stranger's* child. People still tend to save the kid. Apparently we are inclined to favor a child who could grow up and have children of his or her own in the future. The impulse to keep the species going often overrules the impulse to look after your own kin.

This exercise probes the nature and the limits of the drive within us to save members of our family. In evolutionary biology, this drive is called "kin selection." We work hard to preserve or, in this case, save members of our kind to whom we are related, with the nature of the relationship influencing the intensity of the drive. Scientists express it this way—bear with me for a simple mathematical expression:

$$b > c/r$$

(benefit is GREATER THAN the cost divided
by your kinship relationship)

The trick here is the relationship is a fraction, based on the amount of genetic similarity between you and the other person; that fraction is called the r-value. Your own son or daughter comes in with an r-value of ½, because half of his or her genes came from you. Your grandkid comes in at ¼ and so on. So, the closer the relationship, the more benefit you feel you've received for any given effort, i.e. any given cost. This has been observed and documented in nature. But there is a little bit more to it.

Who doesn't love and respect groups of vervet monkeys, in which sisters and aunts and other females help care for each other's infants? (Zoologists really do refer to a "barrel" of monkeys.) Who doesn't love even more the red squirrels who adopt a stranger's buck or doe (baby squirrel) if the mother and father have disappeared? Kin selection, at least for us humans, is easy enough and reasonable to understand in evolutionary terms. When it comes to the burning building, though, this equation doesn't tell the whole story. The fraction on the right-hand side never goes to zero: You, as a human, will tend to save the kid even if he or she is completely unrelated.

You can imagine that families and extended families who nurtured the offspring of their family, i.e. their tribe, barrel, pod (whales), or dray (squirrels) do better, i.e. have more healthier offspring who go on to produce more offspring—children and grandchildren—than those who don't nurture so well. Even schools of certain shrimp protect their young as a group effort.

It is fascinating to note that bees do it. So do ants and termites. In these enormous swarms of individuals an overwhelming majority of them do not reproduce. They were born as offspring of a queen and her few male drones. Biologists reason that, because they share so many genes with the queen, they are driven by motivations deep within their DNA to behave in a way that ensures, to the extent possible, the continuation of their swarm. You may have read accounts or seen videos of the ants that stop at nothing to advance through a forest. You may have encountered a potentially deadly swarm of bees, each of whom will kill herself to sting you and drive you away. It's suicide for the greater good. What else could it be that drives these behaviors other than the creatures' relationships to one another? It's natural selection driven by kinship. Evolutionary biologists call it kin selection.

Biologists have often written about how all this game theory and kin selection plays out among our fellow humans. I know that, as an uncle, I would do just about anything to preserve the lives and well-being of my nieces and nephews, and especially of their offspring, my grandnephews and grandnieces. It's a crazy feeling. But, I know that feeling is there, and I cannot work my way around it. That's not a bad thing, I suppose. It's deep within me and all of us. This same sort of avuncular and tauntular (my own coinage for having the characteristics of or pertaining to an aunt—it's a word strangely missing from the English language) behavior is observed in a pride of lions, for example. Uncle lions often become beta-males and are subservient but helpful all at once. Even though they are not the parent, they are deeply genetically invested. Such is life.

It goes even beyond that. As the thought experiment extends to include our human family, our kin-selection choices expand to include the whole of humanity, as happens to those who imagine saving a stranger's child over their own mother, even though the child's r-value to you is very small indeed. At some point, all of humankind seems like kin. This unfortunate mom isn't going to have any more kids, but that child might. We owe ultimate allegiance to the survival of the species.

This is not just a hypothetical situation. As I write, there was a recent, highly celebrated incident in which a firefighter in Houston, Texas, risked his life by directing his fire truck ladder to set right up against an enormous apartment-house fire. The building was unfinished, and the wind carried the flames through the open-frame lumber like a hot knife through soft butter. The firefighter got right up next to the building, which allowed a construction crewman to jump from the flaming unfinished building to the ladder. A moment later the building collapsed in a flaming hot mess, missing the two men by millimeters. The incident was caught on video.

All of North America watched as one of these two unrelated

guys risked his life to save a stranger. We celebrated their bravery and especially the selflessness of the firefighter. Newscasters speculated as to whether or not they'd ever have reason to sit down together for a beer or just go their separate ways. There was no question in any viewer's mind whether the firefighter and his team on the truck below did the right thing. Of course they did. Saving another life is what we do, whether we're related or not. Thinking a little further along this line, one can imagine that those among us who would not save another's life are generally unwelcome. They are, we might assume, less likely to meet a husband or wife and produce kids to carry those miserable genes into the future.

Altruism is not a moral or religious ideal, no matter what some people might tell you. It is an essential, biological part of who or what we are as a species.

29

COSTLY SIGNALS

There's an old saying around the gym, among the guys who work out and lift weights: "More curls; more girls." For those of you unfamiliar with this time-honored advice, it refers to doing bicep curls with weights to build up a man's arms for no other purpose than to be attractive to women. Jesse Rafalski, a professional trainer, directs his clients on how many repetitions one should do with each weight and in what configuration. He may be the best trainer in Hollywood, and he offers this wisdom: "No matter what they say or tell you, women like muscles." My sister Susan is a student of these things and assures me that it's absolutely true. It seems that Darwin's concept of sexual selection is alive and well and living in gyms across North America.

Really, that should come as no surprise. Our species is prone to it as much as any other that reproduces sexually. Human males seek out females who appear to be highly fertile, which is a likely reason why human females have permanently enlarged breasts—unlike almost all other mammals. Human females seek human males that, according to what they can assess, are well suited to raising children. He has to be a good provider and protector, one worthy of a female's trust. That said, bump into him, then give a quick squeeze. To wit, check out his arms first.

Perhaps this interest in bodybuilding is left over from our ancestors' early days on the savannah. The modern man living in a city (as more than half the world's population now does) will seldom have to lift anything especially heavy. He will not have to defend his home or cave, or his mate, from dangers akin to lions, and tigers, and bears. Instead, he will have to be smart enough and diligent enough to compete with other men for females. Biceps curls are generally no longer an essential part of being a good protector, and certainly not of being a good provider. But there is a way in which biceps are still relevant. The reason that men work out today may not be to send signals to females; it may be to send signals to other men. These signals come at a cost in time, effort, energy, and resources. If you've ever watched the guys at the gym, you know that some of them spend a *lot* of resources.

By appearing strong, an animal sends a signal to enemies and to members of his or her own species, "I am dominant. I am worthy of respect. Do not mess with me!" I am paraphrasing a bit. The important idea is that it costs something to send signals. It takes time to do biceps curls. It takes resources and practice to wear high heels. Perhaps more objectively from a human scientist's point of view, it takes energy and athleticism to pronk, if you're a springbok. *Pronk* is a South African verb that means "to jump vertically." It's a behavior common among the diminutive African antelope, and is a good example of a signaling strategy used by an animal other than us.

On the savannah in Africa, we can observe a dramatic conflict of predator and prey. The big cats like lions and cheetahs hunt foragers like zebra and springbok (which is also the name of a South African rugby football team). From time to time, when a springbok senses the presence of a big deadly cat, a springbok will jump straight up in the air, quite high. This takes energy. It could wear a springbok feller out.

Why would an animal that might require every last bit of energy and concentration to escape death by lion spend any energy or time bouncing up and down? Apparently, to send a signal to any nearby predators that this particular springbok is fit and ready to flee. If you're a big cat, you have to add that into your attack plans and tactics. Keep in mind that the overwhelming majority of big-cat attacks fail. Most of the time the cats get outrun or outmaneuvered as they stalk and pursue prey. It's a matter of life and death for both animals involved. The cats cannot afford to waste energy any more than the springbok can. Jumping around for no reason in an imminent combat situation seems ill advised, unless it has a costly signaling purpose.

Consider the peacock, the epitome of costly signaling. These guys walk around—and more surprising to me as an engineer, these guys *fly* around—carrying enormous tails. From a sexual selection standpoint, the big brightly colored tail is understandable. The males display the tail to attract attention from peahens. They display the tails in billboard fashion to show a peahen of interest that they are healthy and free from parasites. But, the display seems to serve other, related purposes as well. It shows the peahens that the peacock is strong. And it shows other males that this peacock can outstrut, outfly, outcourt, and generally outcompete them. It shows predators that this peacock can show the tail, hit on the peahens, and still fly away if you come looking for trouble. "You got a problem with that, predator? 'Cuz, I'll give you a problem . . ." Once again, I paraphrase.

A crucial element of costly signaling is that the cost has to be obviously real. Waste is an essential feature of signaling to a predator or a potential mate. The springbok can't be faking it. Let's say our springbok is in great shape. He or she can jump well right now. A lioness spies the springbok and considers quietly stalking her way in for the kill. Our springbok sees the lioness and pronks expertly; he executes a good high jump and sticks the landing (as we say in gymnastics). The lioness sees that the springbok is fit and probably can't be run down and eaten; she stalks off, leaving the springbok alone.

Now let's suppose that the lioness comes upon another springbok and wonders if perhaps she can grab this one. But in this case, the springbok pronks only half as high as it can—or more important, only half as high as the lioness expects a springbok to jump. She might conclude that the springbok is injured, that it's not up to a full jump and therefore might be a suitable target for an attack. There could be strategy behind this feeble pronk, however. The springbok might be perfectly healthy, but pronk halfheartedly in an effort to entice the lioness to stalk it, which would give his or her vulnerable calf time to slip away. If this is the case, the hero of the story is taking a big chance. The lioness might move more quickly than the feinting springbok thought, and he or she could easily end up getting caught, killed, and eaten.

The thing to notice in these scenarios (which are uncomplicated by variations in the terrain, the wind direction, and other environmental factors) is that our hero the springbok can pronk all he wants. If the lion thinks she can catch the antelope and she's right, the end result is death. In other words, the springbok has to be able to make good on its ability to escape. He or she has to prove that it can outrun or outdodge the lioness and get away. In similar fashion, the lioness has to be able to catch the occasional springbok, or else springboks would not concern themselves with lions.

Perhaps most important, the halfhearted pronk has to be ambiguous enough to introduce doubt in the lioness's mind so that she pauses. The in-between scenario has to really be in between guaranteed success for either predator or escapee.

This in-between version leads me to consider and embrace what's called the handicap principle in evolutionary biology. It means that the feint or the threat must have a clear cost to the organism making the statement about his or her fitness. If springboks could jump effortlessly all day, lions would recognize that their antics had little or no bearing on their fitness as escape artists. The hunter would not be fooled. But near as we can tell, the pronking is tiring. It takes a little bit out of the springbok, and the lioness knows this. She senses that it's hard for the springbok. As a result, she recognizes that if a springbok can jump that high and show off at that level, that springbok is essentially uncatchable. After millennia of these interactions, the deadly interplay is held in a perfect balance.

In the case of predator and prey on the wide-open savannah, we're talking about physical abilities. But the same process unfolds in many other ways right here in human society. You know the old saying, "Keeping up with the Joneses." We send each other signals continually. We acquire titles and material goods to let others know our value or status relative to everyone in our tribe. We can all understand the urge. It's a human version of peacock feathers, or moose antlers, or the giant neck frill of the Triceratops, or any of a million other types of costly signaling that animals do or did.

If someone else has acquired a new phone or lawn mower that clearly performs better than the one you're using, you want it. From there, it's a small step to feel the urge to be out in front, to make other people want what you've acquired. It costs you something, but you're sending a signal—an evolutionary signal—flashing across deep time.

30

GENETICALLY MODIFIED FOODS—WHAT THE GMF?

If you ever visit Salt Lake City, Utah, I encourage you to drive to Temple Square; it's on North Main just south of Temple Street. There you will see the Seagull Monument, a bronze sculpture atop a granite column commemorating what certain believers call the Miracle of the Gulls. In 1848, Utah farms were attacked by hoards of katydids, insects related to grasshoppers. A group of California gulls showed up and reportedly ate a great many of the katydids. Locals, who were generally of the Mormon religious sect, considered it a miracle. The gulls are the Utah state bird, and there was nothing extraordinary about their presence there, but people took this katydid business very seriously because the threat from insects was so great. By some accounts, the Mormons stood to lose a year's supply of food in less than a week. They built a monument, after all.

Today, farmers cede an average of 13 percent of their crops to insects. It is a serious, serious business. In the 1990s diligent biologists addressed this problem in an evolution-based dramatic new way. Researchers developed ways to extract genes from one species and insert them into the genetic code of another—a technique that can, among other things, cause invading insects to kill themselves.

The organisms produced in this fashion, which we eat, are called Genetically Modified Organisms or GMOs, sometimes Genetically Engineered Organisms (GEOs) or, for this chapter, Genetically Modified Food (GMF). These days, the term is probably most familiar from the "Non GMO" labels slapped on many organic foods, inspired by fears that genetic modification makes things unsafe to eat.

Genetic engineering is a kind of artificial selection, but it's quite a bit different from the breeding traditionally done by farmers, arborists, and dog fanciers. GMOs end up carrying combinations of genes that would not naturally occur inside the cells of any organism. Scientists armed with a rich understanding of the DNA molecule have managed to combine genes from species that would never cross paths in nature, species that could never even be successfully "crossed" in the Darwinian sense by human breeders.

We've been heading down this path for a long time. Have you ever encountered a clone? You may think not, but I'd be surprised if you haven't seen one and enjoyed its fruits. Almost all the strawberries in the world are grown from plants that farmers cultivate by letting the shoots or runners (rhizomes, technically) grow out from a parent plant. The plants that take root from these runners are genetically identical to the parent. Such offspring are called clones, which comes from the Greek word for "twigs." Most grape vines are cultivated in a similar way. Wines are therefore the product of this kind of cloning. It's not an inherently bad thing, especially if you enjoy a good Cabernet. It's just that these days we are able to produce clones artificially, without a parent or predecessor as such. This is another area in which science is wresting control of the genome away from natural evolution.

You may have heard of Dolly the sheep, the first artificially cloned animal. She was produced back in 1996 by putting the DNA from a single cell of one sheep into the egg or ovum of another sheep. That artificially fertilized egg was then implanted in the uterus of a ewe who gave birth to Dolly. In other words, Dolly was genetically identical to her mother, or at least she had exactly the same sequence of genes in her DNA as her genetic mother. Her genes were independent and unconnected to the ewe who gave birth to Dolly. That sheep was what we call a surrogate for what would have been Dolly's traditional biological mother. There was some compelling evidence that Dolly seemed to have been born old, in a genetic sense. There were some chemicals on the ends of her chromosomes called "telomeres" that were shorter than researchers would expect in a newborn. Otherwise, Dolly is and was a regular sheep. She gave birth to six lambs over three seasons, and her descendants still frolic in the Scottish hills today.

To produce Dolly, researchers at the Roslin Institute in Scotland used the DNA from just one cell from a mammary gland of her mother. A cell that forms the body of an organism is called a somatic cell. Most of your cells are somatic; the only exceptions are stem cells and reproductive cells. They say that Dolly the sheep was named for Dolly Parton, who along with a wonderful voice and excellent musical artistry has memorable mammary glands. (I am not kidding; that's really what the Roslin scientists told reporters.) The process was mechanical. The researchers used a long, amazingly thin pipette to poke the DNA into the egg. It was not an easy thing to do. Dolly is the result of implantation number 277, a success that was preceded by 276 failures.

Although it is quite difficult, this type of genetic modification or engineering is, for me at least, relatively easy to understand. We take DNA from one organism and physically or mechanically put it in another. They were both sheep, both the same kind of organism.

But there is a more subtle kind of genetic manipulation that researchers have developed in recent years. This is the work that has led to the creation of GMOs.

There is a way to insert genes into organisms without fertilizing, prodding, or poking: Genetic engineers turn to viruses to do the work. There are certain viruses known to infect plants, for example. Scientists use viruses like these to introduce genetic traits of their choosing. First they splice a desired gene into a virus. Then they infect the plant with the engineered virus, which inserts the gene into the plant's DNA. This technique has been widely applied to corn, soybeans, canola, squash, sugar beets, cotton, and papayas. Along the way, GMOs have sparked controversy for a couple of different and important reasons, one having to do with business, the other with evolution.

The evolutionary concerns revolve around the effects of introducing fundamentally new types of crops into the environment. For instance, scientists have been able to isolate the gene from a bacterium called *Bacillus thuringiensis*, which occurs naturally in the soil around corn plants, and inject it into the corn itself. The reengineered crop, called Bt corn, then produces a chemical that paralyzes the digestive system of European corn-borer caterpillars, causing them to starve to death. Everyone you've ever met has probably eaten food that carries this gene, a gene previously restricted to the bacterial world. About 90 percent of the corn and 93 percent of the soybeans now consumed in the United States are a product of genetic engineering; about 70 percent of processed food includes GMO ingredients.

Researchers have been able to produce varieties of soybeans, canola beans, and corn that make the plants able to thrive even in the presence of the deadly Roundup brand of herbicide. The agricultural seed suppliers charge a good deal extra for this feature, which dramatically increases a farm's yield. If those plants produce seeds that blow into another farmer's field, does the farmer have to pay the seed company a licensing fee? These are the kinds of issues that agricultural businesses are grappling with today.

There is no question that the genetic modification of crops has enabled farmers to feed a great many more people using just a little bit more land. The reduction of losses to pests makes for an almost 30 percent increase in farm yields compared to a century ago. There is controversy about the numbers, but in principle, corn that produces its own insecticide allows farmers to spray less toxic chemicals on their fields. On those grounds, genetically modifying crops seems like a great thing. But many people are opposed to the practice all the same.

Genetically engineering food is controversial, as it should be. If you're asking me, we should stop introducing genes from one species into another, while at the same time taking full advantage of our ability to understand the genome of any organism—plant, animal, or fungus—in order to produce the healthiest, most sustainable food system possible. Here's why: Although we can know exactly what happens to any organism we modify, we just can't quite know what will happen to other species in that modified organism's ecosystem. For me this is a big deal, though some other investigators don't seem to find it as troubling.

Let's look at one well-studied example. Monarch butterflies make an extraordinary four-thousand-kilometer trip from Canada to Mexico every year. When they emerge as caterpillars, they feed on milkweed leaves. (I'm sure they're delicious—the milkweed leaves.) The herbicide Roundup, used by corn farmers to improve productivity by killing weeds, also kills milkweed plants. Genetically modified corn that resists Roundup has also been developed, which enables corn farmers to use more of the herbicide instead of tilling their fields. That practice seems to be inadvertently eliminating large areas of potential Monarch butterfly breeding habitats. There's still controversy about whether the use of Roundup is actually reducing Monarch butterfly populations.

Wait . . . there's a little bit more! There have been some claims (controversial ones, admittedly) that when pollen from genetically modified Bt corn blows onto the milkweed, it can make Monarch caterpillars sick. If that's true, then the genetically modified plants are coming after the Monarchs in two ways at once. What is an industrial farming society—one with a deep understanding of evolution—feeding millions of people around the world, supposed to do? Certain modifications allow us to use smaller amounts of pesticides to raise crops. Producing a sufficient supply of food is an urgent need. It can be a tricky issue. Nevertheless, I still like to evaluate GMOs with the ecosystem in mind.

Consider yet another example. Papaya grown in Hawaii are susceptible to a virus that causes dark circular spots to appear on the fruit. It is appropriately called the "ringspot virus," virus family *Potyviridae.* Researchers found that by putting a piece or section of the virus's DNA into the papaya's DNA, the papayas are no longer susceptible to infection by that particular virus. You don't have to know anything about papaya or fruit or GMOs or farming to tell the difference between a ringspot-infected papaya and a healthy one. There is no question in your human animal mind of which one looks like it would be better to eat.

But there's a little more to the story. There is evidence that by producing a papaya fruit that is not susceptible to ringspot, the papaya plant becomes somewhat more susceptible to another parasite called the blackspot fungus. Furthermore, there is apparently some evidence that some people are allergic to the genetically modified papaya, while not being allergic to the fruit of the premodified, natural plants. In other words, it's complicated. That's the way things are when you change the inputs in an ecosystem. Evolution itself is complicated.

There are also cultural and economic reasons for caution. Case in point: Researchers successfully developed a certain breed of tomato that was far less susceptible to freezing on a cold morning than traditional tomatoes. They did it by arranging for genes from a fish, the winter flounder, to be inserted into tomato genes. These fish, which are pretty common off the coast of the U.S. and Canada, can handle very cold water. The fish gene helped the tomato handle very cold air. That in itself is remarkable, a tribute to our deep scientific understanding of DNA and the effects it can produce for an organism's interaction with the environment. But there is something weird and unnatural about putting fish genes in fruit, in tomatoes. Nobody wanted it, so that research was abandoned.

I'll grant you, this could be a visceral reaction from ignorant consumers. Emotional responses do not necessarily reflect scientific reality, as is evident in everything from creationism to the anti-vaccine movement. In this case, though, I think science and emotion are on the same side. There are very valid scientific reasons to approach GMOs with caution, and those turn out to dovetail with economic reasons. So far, it's not clear that investment in GMOs pays off. It's certainly not clear that GMO research should be funded with tax dollars. If nothing else, the ubiquity of corn and corn syrup has helped to create the weird situation in the developed world, the U.S. especially, of fat people who are malnourished.

Here's one example of how science and economics are getting intertwined. No matter how you feel about fast food, no matter how focused you may be on avoiding fats and carbohydrates, French fries are just delicious. The fries are what keeps me interested in McDonald's restaurants, and I know I'm hardly alone in this abiding interest. McDonald's investigated the possibility of sponsoring research into a genetically modified potato that might make their French fries better and more economical to produce. This was the NewLeaf potato. The company surveyed its customers, asking essentially, "If our fries were even better than they are now, but produced from genetically modified potatoes, would you want them, or even want them all the more?" The answer was a resounding "No!" So McDonald's, the world's largest fast-food restaurant chain, decided not to fund research into GMO potatoes.

That decision essentially killed the research into messing with potato genes—for a while. As I write, there is a new McDonald's GMO controversy, this time surrounding a different modification to potato genes. Certain genes that may lead to the production of a carcinogen when cooked would be "silenced," or turned off, via genetic engineering. Much like last time, the genetic modifiers believe their potatoes are safe to eat. In my opinion we still can't be sure what will happen to the ecosystems out there. The possible consequences give me great pause.

For me, there are two important lessons from humankind's recent experiences with GMOs. They may be good in the short run for upping our food production, and they may make certain foods available to people who would otherwise never enjoy or benefit from those foods, but we don't know—we cannot know—the big picture. What I mean is we *can* determine with great confidence what will happen to the modified organism, i.e. the corncob, the soybean, the canola bean, the papaya, and the tomato. But, we just *cannot* be certain what effect a GMO will have on the ecosystem. We just can't know what will happen to populations of butterflies or even perhaps a population of bats that eats butterflies or the population of fleas that carries the odd bacterium that keeps the population of bats in check, and so on, and so on.

Ecosystems come to be by means of the bottom-up nature of evolution. Natural systems come into existence over thousands of centuries; they end up being extraordinarily complicated. By introducing organisms designed in top-down fashion, there is just a very good (or bad) chance that the designers, the gene-modifying scientists and engineers, will miss something—something subtle but important. So for me, evolutionary theory informs our decisions about GMOs.

In general, natural systems are just too complicated for us to predict the effects of taking genes from one species to another. Instead, we should focus our food-production modifications on those within a species. Hybridizing wheat is great, so long as we do it wheat-to-wheat, working within the framework that evolution has tested for billions of years.

There's another important aspect of the GMO issue for me. When we clear away millions of hectares or acres of land and inundate them with pesticides, the environment is harmed—but the damage is potentially reversible. If we stop using these chemicals, the ecosystem will probably recover. Would that be the case if we introduced a new species that would not have come into existence by natural or hybrid-style breeding? Would nature heal from the negative effects such an organism might have on the environment? It's hard to say. I'd prefer to err on the safe side—not because I'm anti-corporation or anti-progress (not at all)—but because I recognize there's just no way to predict an outcome.

What of the GMOs that are already extant? Well, we'll live with them. They are already being integrated naturally into their ecosystems, even their human-made artificially created ecosystems. Time and the process of evolution will sort out any good or bad ramifications of the food's genes; there's no obvious way to recall them anyway. Meanwhile, instead of continuing our pursuit of extraordinary genes to create extraordinary quantities of food—more than we need in the developed world—let's optimize our farming practices to bring healthier foods to all of us, all over the world.

We have enough good food. We just need to find better ways to bring more healthy food to everyone. If we are choosing our battles, let's pick this one.

31

HUMAN CLONING—NOT COOL

If genetic engineering is confounding when applied to crops, it is downright head-spinning when we think about applying it to people. We now have the capability of bypassing natural selection and imposing a very precise form of artificial selection on ourselves. But it goes further than that: Scientists know enough about human DNA that they could, in principle, clone a human being and bypass the entire billion-year evolution of sex. There are a lot of reasons to be excited about these advances, but let me say up front: Human cloning ain't one of them.

Oh I get it. Most of us wish we could do a lot more every day: more chores, more shopping, more writing, more work, more exercise, all of it. To that end, it's a popular theme right now to propose that we clone ourselves. (I find it funny that people who say they don't believe in evolution are often opposed to cloning—not because they doubt it will work, but because they fear it will work too well.) But if you've ever given birth, witnessed a live birth, or simply watched a little film about it, you'll appreciate that cloning people is somewhat harder than it's depicted in ads for financial planning and family vans. Not only is it difficult, it's not what any of us really wants. Cloning doesn't give you a perfect copy, nor is the process instantaneous.

In the plant world, cloning is easy. Think of all those cloned strawberries and grapes. Bananas and potatoes are grown from clones as well. Heck, you can go online and instantly find a DIY guide of how to clone a cannabis plant. Cloning a mammal, any mammal, is a different animal. But starting with Dolly the sheep in 1996, scientists figured out how to do that, too. They've managed to bypass evolution and its stubborn preference for sexual reproduction. Cloning eliminates the variations that are the raw material of natural selection, replacing them with perfect genetic predictability . . . in principle, at least. At this point researchers have cloned about two-dozen different species. Nobody has cloned a human, yet—at least, no one has admitted to it—but the process would surely work the same way as it did with Dolly.

First you take a cell from one animal and extract its DNA. Then you insert the DNA into the egg or ovum of another animal. If the process works—and a lot of the time it doesn't—the egg takes in the DNA and resets as if it has been fertilized. Put that fertilized egg into an appropriate host, wait through a standard gestation period, and at birth time out pops your clone—your brand-new, crying baby clone. Then you spend a couple decades raising the clone to adulthood. Imagine the petulant teenage years: "I didn't ask to be cloned!"

Contrast that result with the clones in our current commercials, or in a long line of sci-fi movies. In fiction, the clone usually pops out as a fully formed adult. You want more of yourself, so you just somehow make more adults. The real world doesn't work that way. Nobody knows how to make a baby grow into an adult any faster than usual. Then there's also the little issue of nature versus nurture. How could you possibly raise, say, four kids to have exactly the same experiences? Ask anyone who's raised twins, or even met twins. Even more puzzling, how could you raise your clone to be just like yourself? We are all shaped not only by our genes, but also by what happens to us.

Now consider what the fate of a cloned person might be. In our scenario, a person who thought that he or she was a big deal got her or himself cloned. Somebody somewhere was, in this story, able to extract DNA from one of the person's cells, and implant it in a surrogate mother's womb. (I've read articles about women carrying other people's babies for around the price of a luxury automobile.) When this baby is born, he or she would be one genetic step behind his or her contemporaries. He or she would not have the genetic benefit of a new mix of genes, as other (non-cloned) organisms do. The clone would have sidestepped the evolutionary mechanism of sexual reproduction.

This idea that you fall behind when you clone is important. You fall behind in time, in genetics, and in evolution. If people were to stop and just think about that, nobody would even be thinking seriously about making a human clone. Then our lawmakers could relax the controversial laws against cloning research in the United States, and get on with other business. This would, in turn, enable United States' medical researchers to do some basic investigations into the nature of cell genesis, which may lead to new therapies that would improve the quality of life for everyone everywhere.

There are people who object to messing around with human eggs and sperm in any way, based generally on their interpretations of *The Bible*. There is, for example, the strong belief that life begins at the moment a human egg accepts a human sperm and is thereby fertilized. But that's not exactly what happens, or more accurately what has to happen. Once the egg has accepted a sperm and its Y-shaped or X-shaped chromosome, it has to attach itself to the wall of the female's uterus. If it doesn't do that, there will be no baby in the works. After attaching to the uterine wall, the fertilized egg forms a cup shape and three layers. This is "gastrulation" because, if you use your imagination, the cup shape reminds you of part of an intestine. (This business of medical terms based on Greek and Latin is part of why it takes four years to get through medical school.)

Not to put too fine a point on it, but no one in any church would even be able to assert that an egg is viable or not were it not for the scientists with microscopes who studied the details of human eggs and the fertilization process. Certain church professionals go on to claim that they know what an egg does after it's fertilized; this may be a case of a little knowledge being a dangerous thing.

I mention this because our lawmakers spend a great deal of time in town hall meetings and on our legislature debate floors considering laws based on the idea that fertilized eggs are the same as people. Entire branches of organized religion seem to be based on the notion. Violent internecine wars have been fought over the assertion that un-gastrulated eggs are people. It's not clear that they are.

I hope it gives some of us pause for thought to realize that fertilized eggs pass right through women into the environment all the time. Put bluntly, un-gastrulated fertilized eggs become sewage. Are the women who produced these cells to be prosecuted for violating some church-driven law? Are they to be tried for involuntary infanticide? What about their husbands, whose sperm perhaps were not active enough to git 'er done? The science is clear; certain church-derived ethics reflect an understanding that's murky at best and just plain ignorant at worst. Perhaps we should be prosecuting people who espouse these views for undermining our economy, as we will surely send medical patients overseas to spend enormous sums of money elsewhere, to places where certain egg-based treatments are available.

The understanding of processes that lead to babies and eventual adolescent algebra students came from basic scientific research. It did not come from ancient texts or scripture. Without the basic research, this odd debate and these extraordinary laws that are intended to legislate or control what goes on in a woman's womb would not be possible. We have debates that are based on centuries-old scientific discoveries. Perhaps a more informed approach would obviate the need for the debates in the first place.

There is a fundamental difference between research based on fertilized eggs and human cloning, even though the two issues often get lumped together. But making that distinction, and explaining why human cloning is a bad idea, is tricky in a country where many people still close their eyes to the lessons learned from evolution. It's been difficult to get even our elected leaders to take the time to grasp the issues at hand. We'll see in the coming years if learning the facts helps our leaders make informed ethical decisions.

There is a great deal more to the ethics or idea of human cloning. The technique frankly holds great promise, but it also may be just too invasive or just plain weird for many of us. Please, consider the following: In recent years, medical researchers have discovered and investigated the significance of stem cells. These are the cells in a fertilized mammal's egg or ovum that divide and divide and become every single cell in your whole body. People often make reference to the "miracle" of birth. Well, it may not be miraculous by nature's standard. After all, animals have been doing it routinely for hundreds of millions of years; but it is absolutely amazing to me all the same.

Once an egg has been fertilized and managed to attach itself to the wall of the uterus and gastrulates (produces the three layers), the next key stage is the development of the "blastocyst," the sack or sphere of just one hundred fifty cells that is produced by the first few divisions of the single cell that is the fertilized ovum. These cells will divide and divide to become a porpoise, a possum, or a person.

With this understanding of the self-dividing and self-organizing nature of stem cells in the blastocyst, it's not unreasonable to wonder: If a skilled researcher could extract a stem cell, couldn't he or she induce new or regenerative growth in a person who needed a new or replacement organ, for example? This may sound a little creepy, at least at first, but people have proposed doing just that by harvesting an egg and fertilizing it in a laboratory setting, as we do now for in vitro fertilization. Skilled technicians would let the egg divide for the better part of two days and extract the stem cells. These would then be used to help a car crash victim regrow spinal nerve cells. With the aid of such cells, a person with a catastrophic spinal injury might be able to induce his or her own body to grow its own new nerves, and regain the ability to walk. As odd or invasive as it might sound, compare it to cutting people open to put in new heart valves or titanium hips. Those are extraordinarily invasive procedures that have become commonplace in the developed world.

When it comes to the deliberate use of human eggs, they are harvested or extracted all the time, and discarded all the time; that, too, is a part of in vitro fertilization. There are many deeply religious people who object to this. For me, their reasons are arbitrary. An extraordinary number of human eggs go unfertilized—countless unproductive eggs are shed by every single woman on Earth today, and everyone who has ever lived. Imagine a religious imperative to ensure that every dandelion seed reach flower-hood. We'd have a weedy world on our hands. I am just reminding us all that unfertilized eggs, unproductive seeds, and unmet potential are part of the bigger picture of reproduction and, for better or for worse, are the way of the world. This fundamental insight that living things produce a surfeit of eggs and sperm, more than can survive, goes back to Darwin's work on competing populations and figures prominently in any understanding of biology and evolution. For me, this makes every baby that much more precious—*after* the egg successfully develops and a baby is born, not *before* the egg even attaches to its mom. My point of view differs from many other people's, because it's based on the facts of life rather than some suppositions of life.

In 2005, my staff and I were in the laboratory of Hans Keirstead (then at UC Irvine), where laboratory rats are sedated and purposefully injured. Their backs are broken, and when they awaken from anesthesia, they are partially paralyzed. This is the kind of medical research that many of us find troubling. But read on, because the results are astonishing. Sometime after the fracturing impact, the spines of these rats are injected with human stem cells. These have been grown from fewer than ten strains of stem cells that date back decades, when stem cell research was regulated differently. These stem cell lines have been continued and nurtured in exceptional, carefully controlled laboratories. Along with the human stem cells, the rats are given the same type of drugs used to prevent organ recipients' bodies from rejecting the donor's organ.

In a few days, the rats are able to move their back legs. In a few weeks, they are walking pretty well, and regain control of their bowels. That is, they were incontinent after the spinal injury (even for rats, it can be trouble). Upon close examination, the nerves in the spines of the rats have grown back, to a large extent. They are regenerating their own spines in their own backs. It is amazing. The ramifications for human medicine are far-reaching. Or they could be.

Here's the thing: The technology to extract stem cells from a fertilized and growing human egg is exactly the same technology, at least right now, as the means by which we can implant somatic cell DNA into an egg to start the process of cloning. It is a line of reasoning that we should all follow and form an opinion on, because as the techniques become more refined in animal research, the possibility of performing humane work becomes feasible. Does it become imperative?

How do we draw a clear line that allows therapeutic work but prevents full-blown human cloning? That's for all of us taxpayers and voters to ponder.

Because a blastocyst would be involved, researchers are looking to isolate and grow stem cells not from an egg (which could, in principle, develop into a viable fetus) but from a cell taken from another part of the human body. Perhaps one day soon, regeneration of one's own organ cells or nerve cells will be possible and as common as a hip joint replacement. When you consider the extraordinary potential cost savings in having a patient's body do all the work compared with the costs of high-tech prosthetics, it may be an ethical imperative to use a patient's own stem cells for this sort of improvement in his or her quality of life.

As you think further about this research and the potential ramifications of the use of human stem cells in medicine, keep in mind that research on rats and other animals is possible because we are all so much alike. This is direct evidence of evolution. We share so much with our placental mammalian cousins because we all had a common ancestor around 70 million years ago. Our understanding of medicine, blood types, the central nervous system, and where we ultimately came from are all direct results of our understanding of evolution.

If we did not have a common ancestor, if we did not share DNA sequences, if we had not all descended from ancestral living things, all of life science, all that we see living in nature would be far, far more mysterious and hard to understand. That the essential discovery of evolution was made barely a century and a half ago indicates for me how primitive we all must still be. We are only now starting to use our knowledge of living things, gained through the study of natural selection, to become more compassionate toward our fellow humans and better stewards of Earth. We have a long, long way to go. It's exciting to contemplate.

32

OUR SKIN COLORS

As an elementary-school kid in the 1960s, I was very aware of race and racism. Back then, Washington, D.C., was in many ways a racist Southern town. That awful sentiment was just part of the scene. I could hear it without listening very hard. It was right there in background conversations in busy restaurants, along with the sound of clinking glassware. I could also see what was happening. Newspaper headlines described famous people who were assassinated for clearly racial reasons. At the same time, the civil rights movement achieved sweeping changes in laws and perceptions. It all had a deep effect on me. Today, it's easy for me to see that racism starts on the surface. Nowhere are the forces of evolution that have shaped our societies more apparent than in the color of our skin. The surprising thing is that skin color does not mean what a great many of us think it does.

I've traveled around the world a little bit (largely because I often attend the International Astronautical Congress, which is held in a different city every year). If nothing else, I've learned that people are a great deal more alike than they are different. In evolutionary terms or fact, we are all almost identical. We each share 99.9 percent of the same DNA. I can prove it to you. Better yet, you can prove it to yourself. What do you think would happen if a man from Scandinavia married and enjoyed sexual encounters with a woman from East Africa? They might easily have a child. That kid would be a human. This union is not going to produce anything else but a human.

There is only one species of *Homo sapiens*. We all share common ancestors. This may be where myths like Adam and Eve living together in a garden come from. If you just sit and think about it, and realize that we are all extraordinarily alike, you might just conclude that there must have been an original pair of humans that led to you, me, and everybody we'll ever see. The author or authors of the Book of Genesis may have reached the same conclusion logically, i.e. by just thinking about it. Humans, all of us, must have had a common ancestor. Otherwise, how could we all be of one species, able to reproduce so effectively? There are more than 7 billion of us strutting, sexing, and texting in the world today.

Despite that line of reasoning, people of different tribes or geographic regions have been warring with each other, distrusting each other, and forbidding marriages with each other for millennia. In many instances, skin color has been a trigger for these conflicts. It raises a fascinating evolutionary question: If we are all one species, why do our colors differ so dramatically? Is skin color connected with deeper differences between various groups of humans? Or you could turn it around and ask, are races real?

Here's the short answer: No. Skin color is a tiny, recent, and transient feature of human genetics. One of *The Eyes of Nye* television shows was about this issue. I stood in a field with a few dozen cattle. I pointed out that the colors of those particular animals varied widely. There were black, white, and brown cows of various shades, and they exhibited no discrimination. They roamed and grazed showing no preference for any one variation. Racism did not seem to be an issue with them; they were all of one species. So are we. Race, as it is commonly understood, is an illusion. But don't take my word on it. Let's see what two centuries of evolutionary research have to say.

The first place to look for answers is on the skin of our closest primate relatives. As our understanding of DNA has increased, we have come to understand that we share around 98.8 percent of our gene sequence with chimpanzees. This is striking evidence for chimps and chumps to have a common ancestor. Chimpanzees have very light-colored skin; you can see for yourself where the skin is exposed on their cute cheeks and jowls. So, we might expect humans to have about the same color of skin as our very close genetic cousins. But with few exceptions, we don't.

Anthropologists have been all over the world looking for fossils of our ancestors. And they've found them: dozens of closely related human skulls and nearly-human skulls (my old boss?) and other bones, which all indicate that humankind started out in East Africa. That's where we find the greatest genetic diversity of humans to this day, and that's where we find the oldest fossils and oldest evidence of human activity. If we started out related to an ancestor of chimpanzees, did we start out with skin about the same color? Fossil bones can't tell us. At least, if there is a way to extract information about the color of the skin that once protected those bones we haven't figured it out yet.

Since fossils don't provide the answer, scientists have turned around and tried to understand the adaptive function of skin color from an evolutionary point of view. Most obviously, our skin protects our insides from what comes at us from the outside—wind, rain, and thwipping tree branches. Less obviously, skin is an organ that produces a chemical we cannot live without, vitamin D. This vitamin got its D designation by being the fourth one to be identified. The main form is cholecalciferol; that's $C_{27}H_{44}O$. Researchers studied our beloved dogs to find it. The ability to synthesize vitamin D goes way back in evolution. Plankton at sea have been manufacturing it for 500 million years. Sea creatures use it to capture and make use of calcium in their environment. So do we.

One of the wonderful things about vitamin D is that you don't need to eat it; your body can make it. You and I just need a little exposure to ultraviolet light to give a form of cholesterol in the skin a jolt and convert it to vitamin D. But there is a catch: Too much ultraviolet light means trouble. It carries more energy than visible light and is able to break down or dissociate delicate biological molecules, especially your folic acid, your folates. Ultraviolet can burn your skin, for example. So to be successful, animals like us need a way to block a large fraction of the ultraviolet light that hits us while letting just enough through to maintain a proper level of vitamin D.

If you're a chimpanzee, you have solved this problem with one of the oldest tricks in the evolutionary book. You grow hair. Hair is made of keratin proteins that are not too different from those of distant relatives on the Tree of Life who use keratin to produce feathers and scales. As you no doubt realize, hair blocks light. Chimpanzee bodies are covered almost everywhere with thick, dark hair. Even if you make jokes about hairy guys, the hairiest among us don't come close. Chimps are protected from excessive UV rays by their hair. But if hair offers such good protection, why did we humans lose most of ours?

A likely answer emerges from the following thought experiment, which I hope none of us ever has to actually conduct. You may have heard about people who kept a chimpanzee as a pet. Everything is just fine while the chimp is young, but as he or she grows older, a chimpanzee proves to be surprisingly strong compared to a human. A chimpanzee can easily outwrestle you and pull your arm right out of its socket. But if push comes to shove and you get into a violent disagreement with a chimpanzee, there's one thing to do. You can outrun him or her.

Humans are champion long-distance runners. As soon as a person and a chimp start running they both get hot. Chimps quickly overheat; humans do not, because they are much better at shedding body heat. You see where I'm going. According to one leading theory, ancestral humans lost their hair over successive generations of hunter, gatherer, scavenger offspring because less hair meant cooler, more effective long-distance running. That ability let our ancestors outmaneuver and outrun prey. Try wearing a couple of extra jackets—or better yet, fur coats—on a hot humid day and run a mile. Now, take those jackets off and try it again. You'll see what a difference a lack of fur makes. Overheating slows us endothermic animals down.

With the loss of hair our ancestors faced a new challenge, though, because it exposed them to more ultraviolet light. Ultraviolet helps make vitamin D, yes, but it also breaks down similarly crucial folic acid along with other ultraviolet sensitive compounds. You get your folates (like folic acid) from leafy green vegetables. Unlike the case with vitamin D, the human body cannot manufacture its own folates. Folates play a role in the growth of fetuses. Babies born to women who have been exposed to too much ultraviolet have defects in their spines and central nervous systems. They cannot survive. So those early humans who had a little extra sun blocking melanin in their skin did better in strong ultraviolet light environments than those with paler skin.

In general, the closer people live to the equator, the more ultraviolet exposure they receive and the darker their average skin color. Strong local weather conditions can also attenuate the ultraviolet levels. Take a look at the map of skin color of people native to different regions of Earth. Near the equator, people have darker skin. Where it's cloudy a lot, as it is in Britain, people have lighter skin. Where people live closer to outer space, as they do in Tibet, they are exposed to more ultraviolet and have darker skin. Skin color is basically a measure of the local ultraviolet levels, and it is controlled by relatively minor adaptive changes in the genome.

This fascinating line of reasoning was explained to me by Nina Jablonski, the scientist who did the fundamental research. (She was at the California Academy of Sciences at the time; now she's at Penn State.) She points out that between two people—any two people—the genetic differences are minuscule. Jablonski is a studious observer of those subtle differences. For her, the most stable physical distinctions are not in the color of our skin, but in the configuration of our bones, especially the shape of our heads. During our interview, she paused and leaned toward me with her hands open as though she were going to reach for a melon on a shelf. As she did she remarked, "Bill, you have a perfect European skull." I had to stop her and explain that she couldn't have my skull right then, as I was still using it (I'm *still* very attached to it). But she got my attention for the rest of what she had to say about human migrations around the world.

With their success in Africa, our human ancestors went looking for new pastures and presumably new adventures to the north in what is now Iran and Iraq. Here, farther from the equator than East Africa is, any individuals who happened to be born with slightly lighter skin did a little better. They got just the right amount of vitamin D without breaking down their folic acid and other essential, ultraviolet-sensitive compounds. They survived a little longer than those who were born in this region but retained African, or very dark skin. Sure enough, modern Northern Africans have lighter skin than equatorial East Africans.

Migration toward and away from the equator led very quickly

to changes in our skin color. The remarkable thing is that all people, in all places, evolved similar skin color as they moved to places with similar levels of ultraviolet light. As our ancestors developed agriculture, they moved from Africa to Mesopotamia and then east across Eurasia. If they wandered southward into what is now India, the offspring who had slightly darker skin fared better than those with lighter skin. People native to southern India often have skin so dark it seems almost blue.

Here's the punch line: Just like their ancestors back in Africa, southern Indians' dark skin color comes from melanin, but in southern India the pigment is turned on by a different combination of genes. They seem to have retained some of their ancestral pigmentation from their African ancestors, but have some additional gene combinations that help produce melanin polymers. These people migrated to southern India from an area that receives slightly less ultraviolet exposure, and they chanced on a melanin-activating gene that helped their offspring survive under a bit more ultraviolet. The changes in skin color happened because of completely independent mutations in skin color genes.

A similar pattern emerged when people moved from northeast Asia (low ultraviolet) across the ice-age land bridge that is now the Bering Strait into North America and south into Central and South America (higher ultraviolet). Their skin color is produced by melanin in lockstep with the intensity of ultraviolet light in the land where they live. The same process produced lighter skin color in the groups of people who moved into relatively sunless Western Europe and eastern Asia. They lost pigmentation in order to maximize their ability to make vitamin D. Where it's sunny year round, native people have dark skin. Where it's only seasonally sunny, natives have much lighter skin. It's true everywhere.

This convergent evolution of melanin provides further evidence that skin color cannot be taken as a mark of racial identity. It also provides yet another way to trace human evolution. The existence of two distinct skin-pigment genes indicates that humans migrated out of Africa separately to areas to the north and east, probably at two different times. Then, through convergent evolution, the Indian populations ended up with melanin-triggering genes that gave their babies the same kind of advantage in a high-ultraviolet environment as the other melanin-triggering gene present in the people of East Africa.

When we compare the skin color of people native to Tibet, who live at high altitudes, with those in surrounding countries, we find that the Tibetan people have slightly darker skin than those neighbors at lower altitudes. By the way, they can get a tan easily. That makes sense. When you're at a high altitude, there is less atmosphere between you and the Sun, so you and your neighbors are illuminated with more ultraviolet rays than those living down below. Your offspring who had the slightly darker skin have a better chance of surviving and giving you grandchildren than those who had skin a tad too light for producing a good extended family.

Skin color is so sensitive to environment that scientists can study the skin colors of indigenous peoples to map human migrations out of Africa and across the world. Modern humans, *Homo sapiens*, first left Africa about 80,000 years ago. We moved through Mesopotamia and started across Eurasia starting around 60,000 years ago. Just 15,000 years ago, people crossed the Bering Strait from Siberia to North America. Now take a look at the two maps. As Africans explored to the east and north, native tribes developed lighter and lighter skin in successive generations. As they migrated south, their skin got dark again. Moving farther east and farther north, their skin color changes back to a somewhat lighter shade.

By the time humankind had made its way to South America, the tribes doing well there had developed skin almost as dark as those ancestral tribes back in Africa. But there's a bit more going on here than meets the eye, as Jablonski found when she compiled data from several studies and documented the relationship between ultraviolet, vitamin D, and folic acid.

American tribes living in tropical latitudes do indeed have dark skin, as do those back in Africa. But, the American tribes have skin that is not quite as dark; it's ever-so slightly lighter than Africans. Why might that be? For one thing, people haven't been in the Americas as

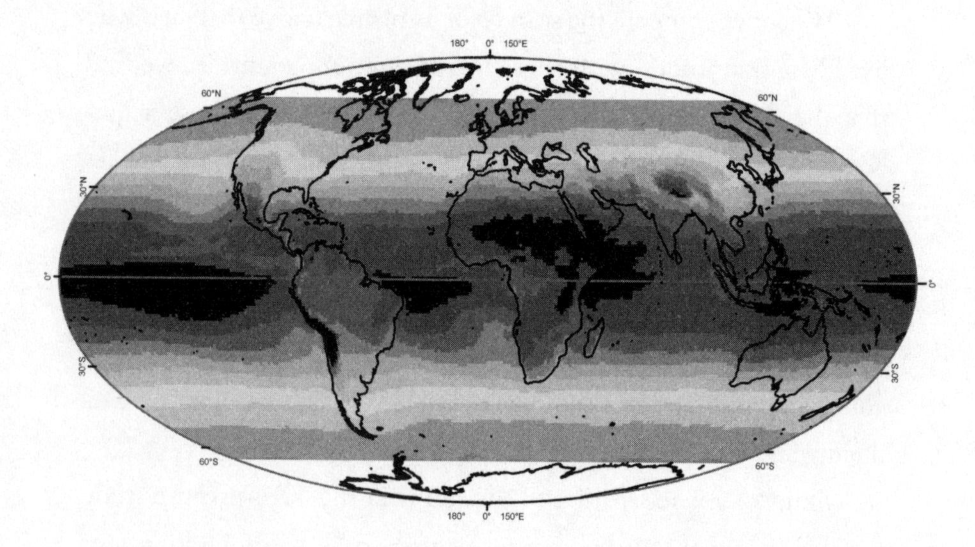

long as they have been in Asia and Africa, so the evolutionary changes have not had as much time to accumulate. Scientists also note that, by the time humans made it south along the American coasts, people had developed straightforward ultraviolet protection technology: I'm talking about hats and jackets. Their tribes had formed the habit of dressing, and that stayed the progression of darker and darker skin as they headed south, where sunshine is strongest. It is just amazing.

The takeaway message here, as Jablonski points out, is that there is no such thing as different races of humans. Any differences we traditionally associate with race are a product of our need for vitamin

D and our relationship to the Sun. Just a few clusters of genes control skin color; the changes in skin color are recent; they've gone back and forth with migrations; they are not the same even among two groups with similarly dark skin; and they are tiny compared to the total human genome. So skin color and "race" are neither significant nor consistent defining traits. We all descended from the same African ancestors, with little genetic separation from each other. The different colors or tones of skin are the result of an evolutionary response to ultraviolet light in local environments. Everybody has brown skin

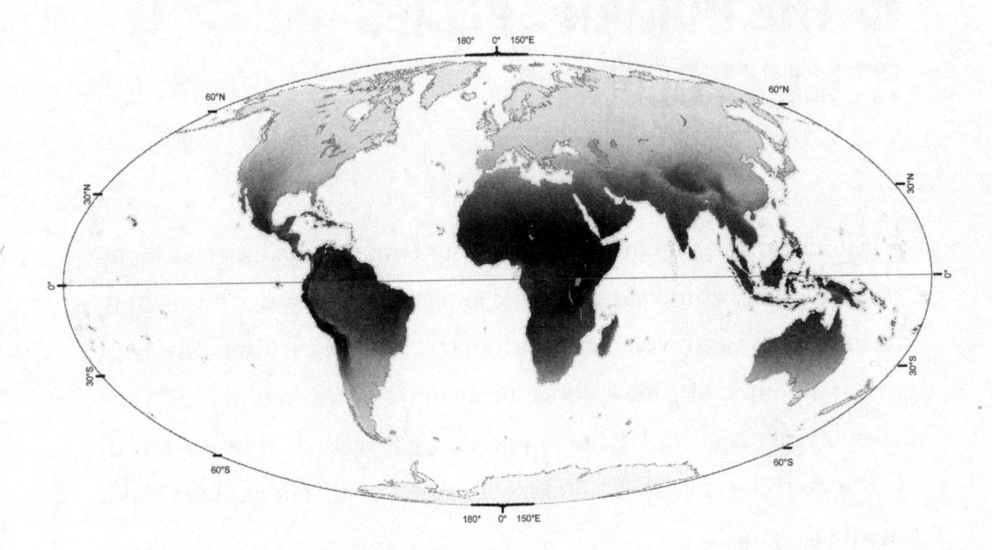

tinted by the pigment melanin. Some people have light brown skin. Some people have dark brown skin. But we all are brown, brown, brown.

Our reactions to other groups are real enough, but evolutionary biology shows that those reactions have nothing to do with race, because race is not real. Scientifically speaking, there is tribalism and group bias, but there cannot be any such thing as racism. We are all one.

33

IS THE HUMAN SPECIES STILL EVOLVING?

I still have strong memories of meeting Ivan. He had a great many fans, so I was behind the barricade and could only wave, but I think he noticed me and waved back. I found it all very exciting. This happened a couple of times—once in Tacoma, Washington, and then several years later in Atlanta, Georgia. I got so excited because Ivan was something special. Unlike most readers of this book, Ivan was a mountain gorilla.

Modern gene sequencing informs us that Ivan and I had a lot in common. Inside, we are about 97 percent the same. It's a short evolutionary distance between us, and as I stood there, I could feel it—so close, yet so far. Why did my line diverge from Ivan's over the last few million years? Looking at Ivan, I couldn't help but wonder how we will continue to change. Will we take control of our own evolution, the way we have taken control of our crops and started taking control of our stem cells? If you were in charge of your own genome, what improvement would you specify? And what about Ivan: Will we be leaving his kind, or primate kin, further and further away on the Tree of Life?

I visited Ivan while I was working at Boeing in the Seattle area, making subtle improvements to the 747 airplane. Ivan was well-known. Going to see him was just something you did as a Pacific Northwest sightseer, like going to the Western Washington State Fair in Puyallup [Pew-AL-upp] for scones and strawberries. At that time, Ivan lived in a concrete box in the basement of this odd old department store called "the B&I," named for its original owners Bradshaw and Irwin. He had a rubber tire swing and a great many bananas. He was brought to the U.S. as a baby around 1964. His sister died young, and Ivan carried on in his box with his swing until he was thirty-two years old.

As I looked through the glass of that weird store basement, it was impossible not to compare the two of us. It's astonishing how closely related we are, yet how different we look and behave. I don't know much about gorillas, but to my human mind, he looked bored with his life, and I could see why. Humans, lots of humans. Same food every day. Rubber tire on chain. By then, the B&I was no longer in its glory days. It was on its way out of business. A deal was made, and Ivan was transferred to the Atlanta Zoo. That's where I saw him last.

By that time, I was working for Disney, frequently passing through Atlanta. I met an old friend and her kids at the Atlanta Zoo. And there was Ivan again. On that visit, Ivan looked great. He now had several girl gorillas clearly smitten with his gorilla-ness. He basked in the sun and interacted with the ladies. I felt deep relief. He seemed a lot happier. By that I mean his posture and the way he moved among his zoo family. In *Julius Caesar*, by William Shakespeare, I remember the dialog, where Cassius says, " 'Tis Cinna, I do know him by his gait; He's a friend." I remember realizing, first of all, you can recognize someone by the way he walks, and furthermore, you can often infer a good deal about that person by his posture and movement. That's how I felt about my buddy Ivan. He was in a bad mood in Tacoma and a very good mood in Atlanta.

For some reason, this kind of meditation about my primate relatives kept following me, though meditation may be an overstatement. I got my start in television on Seattle's local comedy show *Almost Live!* (Bob Nelson, who wrote the Oscar-nominated film *Nebraska*, and the terrific actor Joel McHale started on that show as well.) I impersonated Ivan. Wearing a gorilla suit, and using subtitles, Ivan (well, mock-Ivan) explained that everything was fine for me, really, except I was stuck in Tacoma! In the course of related comedy bits, we had Ivan playing tennis. Anyone watching can tell it's me playing tennis, even though I was dressed in a gorilla suit. You knew me by my gait. I did my best to seem bored and angry, as I presumed Ivan to be. My claim is that a primate's behavior is evident for the same reasons our behavior and posture are evident. It's deep within us. It's in that 97-percent-in-common DNA.

About three years later I was working on my own show, *Bill Nye the Science Guy*. Show number 53 is entitled "Mammals." As part of the filming I visited another gorilla named Vip at the Woodland Park Zoo in Seattle. He was fine. I mean, he was gnawing away on what looked to me like bok choy, and he was looking at me. I felt like I could sense his resentment and resignation. I paraphrase: "What's the deal, man? You're out there with no hair. You look like a pathetic weakling; I feel like bending you over my knee and snapping you in half—and I'm a vegetarian. Somehow, you and all your kind got us all stuck in here, on this side of that see-through wall [thick glass]. I mean, dude—this sucks. And thanks for the lousy bok choy, you hairless geek . . ."

I admit, this dialog was entirely in my head, but I challenge anyone to go to a zoo where you can get close to gorillas, look them in the eyes, and not feel that they know what the deal is. You and I have the natural dumb luck to be more clever and dexterous than the gorillas are. They can't quite tell how the humans set this all up, but they can tell that they got the shorter end of the genetic stick. You have to wonder if this is how we would feel if we could gaze upon future humans—or whatever it is that we become.

Since scientists and engineers developed machines and chemical reagents that can determine the exact sequence of chemical bases in strands of DNA, we have discovered that humans and gorillas share 97 percent of their genetic code. We may be just 3 percent different in our DNA but wow, how different we look and behave (my old boss excluded). From time to time, you hear people make reference to the 800-pound gorilla in the room. Although for most gorillas, 800 pounds would be an overestimate; it's usually closer to 500. Gorillas generally walk with the feet and their knuckles. We're feet exclusive. Gorillas don't quite stand upright. We do. Gorillas are covered with hair. We are not. Gorilla brains are bigger than ours, but ours are larger compared with our weight and size. All those distinctions emerge from that little 3 percent.

What would things be like, if we were just 1.2 percent different? That is roughly the amount of genetic separation between humans and chimpanzees. What about 0.5 percent different? Well, things would still be different. The other creatures would look almost like us. They might be a little heavyset, but they could walk among us and not look out of place. They would probably behave much as we do. They might be stronger but not quite as sharp before a chessboard or algebra problem.

The 0.5 percent gap apparently describes about a dozen human-like creatures that lived here on Earth less than four million years ago. We're talking about the hominids, members of our immediate evolutionary family unearthed by paleontologists. You've undoubtedly heard of *Neanderthals* and *Cro-Magnons*. Are you also familiar with the extraordinary finds at these archaeological sites: *Zhoukoudians,*

Ngandongians, Sangirans, and *Saint-Césaireans*? The hominids who lived there probably had DNA that was 99.89 percent the same as ours, yet here we are, and they're gone—all of them. Somehow, our ancestors outcompeted them all. Maybe our more recent ancestors were better at mapmaking or storytelling, or pattern recognition in general. Maybe our immediate predecessors had a gene that made them just a little more resistant to a certain type of malaria. Everybody else died out except our direct *Homo sapiens* line.

If my buddy Ivan the gorilla was just 3 percent off from us, and we can build so many more amazing weapons than he could, just think what it would be like to meet aliens whose DNA molecules were 3 percent ahead of ours. We'd have no chance. (Of course they might not have DNA, and the difference might be way more than 3 percent, which hardly makes things look better for us.) In the course of natural selection, a tiny difference in DNA could make all the difference in competing for resources, preparing for a hard winter, or just figuring out escape routes before you needed to get away from the stronger, tougher caveman-style tribe that might be pursuing you.

Modern humans are the result of a burst of evolutionary innovation that took place over the past forty thousand years, probably due in part to a bottleneck. It was punctuated equilibrium in action, somewhere in East Africa. A tribe of our ancestors got cut off from the rest. A favorable set of mutations came through, and we've been passing them along ever since. The humans who passed through that bottleneck were nearly uniform genetically, as typically is the case when you are dealing with a fairly small population. We've been changing faster than ever over the past ten thousand years, and probably up through the past few hundred years. We like to think we're immune to evolution, that we've moved beyond it, but we're still in the thick of it. We just can't quite see the forest for the trees.

The human population continues to grow. When I visited the

World's Fair in New York City in 1965, there were 3 billion of us. Today, there are almost 7.2 billion with a b. If there are going to be new gene mutations with new advantages, they'll most likely come from Sub-Saharan Africa, because that's where the most new humans are being produced right now. What will those innovations be exactly? Will the next big innovation in human DNA provide the ability to deal with climate change and high-speed information from all of our devices? Time will tell.

Since all of the other hominids have disappeared, will we be replaced by the next generation of better-built hominids? Is there a *Homo superius* just around the next deep-time corner, waiting to take our place? Let's think about what it would take: If we were to give rise to a new species, something would have to happen to us to create a bottleneck or isolated place for a founder-person and her or his mate to show up and get separated from you and me and our offspring. In the modern world, that is very unlikely. We have airplanes and ships and the Internet.

Circumstances matter; humans are subject to contingency just like any other species. What if we fail to develop defenses to an incoming asteroid? Or if there were an enormous war and all of our intercontinental means of transportation were destroyed? Perhaps then an isolated population of people would live apart from the rest of us for so long that they would no longer be able to successfully interbreed. Just listen to all the dialects we have for speaking English. When human populations are even a little bit separate, they start talking differently. Other bigger changes might happen with more profound separation. Perhaps this could happen somewhere beyond Earth, even, such as in a colony on Mars. Without geographic isolation, I am not sure we can get a new species of hominid, not ever. But that is not the same thing as saying that humans are no longer evolving, because we surely are.

We cannot step away from evolution. Our genomes are always collecting mutations, and we are always making mate selections. Are humans preferentially mating with other humans who are tall? Blonde or not blonde? Sweet, or bitches and jerks? With all of our glamor magazines and self-help books, are we slowly producing offspring that are smarter and better looking? I'll confess that I read the first of the *Fifty Shades of Grey* books. What I got out of it was that an ideal man is a fellow who is young, attractive, and astonishingly wealthy. Who would have guessed? I can't help wondering if that is part of the selection effect that is moving humans steadily further away from Ivan.

Are smart humans choosing other smart humans with whom to have babies, and is that paying off in genetic success? Are they actually producing significantly smarter offspring, who end up making more money and ever so slowly outcompeting other families? Or is intelligence a losing trait, because highly educated couples tend to have smaller families, so when something goes wrong there are fewer siblings left to carry the genes forward? Or since highly educated men and women have babies later in life than those that don't squander their best childbearing years in universities, do the babies of the highly educated enter the world with more trouble in childbirth, and are they prone to more subtle gene troubles that result from later mother and fatherhood? Cue the spooky music.

I am reminded of an old routine by the comedian Steve Martin, who asked: "Do you remember when the world blew up? Remember? We all had to come to this planet on that giant space ark? Remember, the government decided not to tell all the stupid people, because they were afraid that . . ." He let that sentence taper off, while the audience quickly realized they were being made fun of as "the stupid people" in the joke.

More likely than a future race of hyper-smart people who outcompete the rest of us is a strain of *Homo sapiens* that can beat a disease. Probably the most important evolutionary sieve that any future person is going to have to get through is going to have to do with germs and parasites. Recall that in the Spanish Flu epidemic of 1918–1919, some 50 million people were killed by something far too small to even see, let alone hunt and destroy. The Black Death of the fourteenth century may have killed up to 200 million. You and I are descendants of people who just happen to have the genes to fight off deadly viruses and bacteria.

Those who survive into the future will probably have resistance to certain diseases that none of us have today. There are a lot of other ways that evolutionary change will march on, no matter what. Those that survive may have a higher tolerance for drinking milk. Babies in industrialized societies have access to milk like no one before us. Maybe a genetic tolerance for milk will slowly help more of those babies survive until they have kids of their own. There is evidence that people with both especially high and especially low blood sugar levels have fewer offspring. So subtle changes at least will make their way into the human population's gene pool. It's going on right now.

Then there is a whole other category of possible human futures that are influenced by our own technologies. I give a great many talks or lectures at universities and for general audiences. I enjoy performing, the part where I'm doing the talking and all, but my favorite part of any evening is when audience members come up to microphones and ask me questions. One of the most common themes is what people call "The Singularity." This is a supposed imminent time (2029, in some versions of the story) when computers will become as sophisticated as human brains. From there, it is proposed, machines will be able to outcompete humans at just about everything. There will be superior car-parking algorithms, disaster-relief coordination, legal briefs, rocket science, great thinking in general. Taking it the next logical step, this artificial intelligence will have to be managed carefully, because after all, any of these future brain machines will outthink and outmaneuver us at every practical turn.

Look, I love thinking in big ways about the future. I love science fiction as much as the next guy. But, I am skeptical that this singularity will be an especially special moment in human history. I say this because somebody has to provide the power for these machines. Somebody has to figuratively (and until 2050 or beyond, still literally, I suspect) shovel the coal. Here in the 2000-teens, we have more mobile phones than we have humans. Even so, only about half of us have access to those devices. There are places in rural Africa and rural China, where people have never made a phone call of any kind. The singularity won't affect them for some time, if at all.

I'm reminded of the 1970 movie *Colossus: The Forbin Project*, which was based on the 1966 novel *Colossus*. In the story, both the U.S. and the Soviet Union use computers to control their astonishingly large arsenals of nuclear weapons. The U.S. machine, called "Colossus," is in an impenetrable bunker and is powered by its own nuclear reactor (how hard could that be?). Thinking it will help avoid trouble (actually *more* trouble by this point in the story), it's agreed to link the two super machines into one huge machine. As you might imagine, things go wrong in a huge way.

Included in thinking about the singularity is the idea that our information technology will be so sophisticated that we will be able to put human consciousness in some computerlike machine. There are organizations that believe in what they call transhumanism. Humans could then, in a sense, live forever . . . that is, unless someone pulls the plug. A friend of mine wears an ankle bracelet that explains what to do with her head when she's dead. She wants it frozen in the belief or hope that sometime in the future, we'll be able to connect her dead head to the right machine and bring her brain back to life, or somehow download her stored consciousness. One of the Web sites for her head-storage company brags that they've never had a problem with their cryogenic system, and they've been operating since 1976. It's foolproof, because all they have to do is add liquid nitrogen every three weeks. Wait. Where does one get liquid nitrogen? It's produced using electricity from a power plant. In Michigan, where this company is located, the electricity mostly comes from burning coal. All good for the frozen heads till the grid goes down.

Meanwhile, regular people will be having babies, who may find a great many more interesting things to do than converse with dead people in electric brain machines. Most of those babies will be born in the developing world, which is generally far, far away from these extraordinary future brain tech centers. Of course, as I often remark, I may be wrong. The Colossus machines of the future may be designed to efficiently run entire cities, and they may do it perfectly. It's not hard to imagine sewer systems, solar energy systems, and transportation systems all being directed by a big brain of the future. Nevertheless, I believe sex and nurture will be the main way most of us move our genes into the future even after the singularity machine is debugged in the laboratory.

If you want to reach toward a science fiction future of human evolution—it's fun to speculate, so why not?—a much more reasonable, perhaps inevitable, factor is genetic engineering. Medical scientists are already on the verge of being able to ensure that your baby does not suffer from Huntington's disease or have flat arched feet. Will it be possible to make babies genetically smarter? Or better baseball players? With all of this being done in a petri dish to the eggs and sperm before they're fused? Is that sort of thing ethical? More important, if we make smart people, will they be socially comfortable?

The science fiction stories about superior people produced by genetic engineering almost always end badly. The superior people end up causing too much trouble, generally because they don't fit in with the rest of us. In the real world, these issues will unfold incrementally and, very likely, with a lot of controversy. As a voter and taxpayer, each of us may have some interesting decisions to make about what's allowed in medicine. We may have to address genetically modified people among us who came to be through some future outlawed genetic-engineering technique. What would be the status of illicit human clones? The better informed we all are about all this, the better decisions we'll make.

In contrast, I'm looking out for big changes that come from good old-fashioned Darwinian natural selection. What trait would give a future human baby such an edge that she or he will grow up to produce some amazing new kid that can do something that stands out and will attract a similarly worthy partner with whom to mate? I have heard many women say that they love a guy with a good sense of humor. That one sits well with me for some reason. Will some future guy be so funny, and not so funny looking, that his hilarious sense of humor will win him partners? Will his command of irony be so good that women go wild for him and he mates and reproduces wildly? Legions of present and former stand-up comics hope so. (Many among those legions believe it to their core.) Will it be a guy who can develop such fantastic pectoral muscles, which women find so hot, that he mates and mates?

By the way, what is the sexiest thing about a woman? This is not a trick question. It does not require men to cower fearing a politically incorrect secondary remark or facial gesture. In my opinion, the sexiest thing about a woman is her smile. If the woman doesn't smile, or doesn't smile well, men will not dig her. They will look for other women, who smile well. What's involved in smiling? Good teeth, attentiveness, engaging eyes, and the ability to be happy. Each of these is an apparently inheritable trait. Each is not going to be much affected by sleek computers hooked up to a nest of frozen heads.

A smile comes from deep within. If it's not genuine we can tell, albeit not consciously, and often not right away. Will a future generation of women smile wonderfully, because good smilers are more likely to attract a man who can support them as they successfully reproduce, ensuring that both her own and the man's genes get passed into the future? Is it more likely that the woman's genes will be passed on if she's just tougher in childbirth? Or is it just that whatever genes anyone has stand a better chance of going forward if she or he lives in an industrialized society where appendectomies (like mine) are routine?

Another consideration: Any distinctive genetic traits—good or bad—that happen to be in a population with access to effective health care will get passed on. Health care takes away certain selection pressures, and may introduce others. Unselected genes get passed on generally because that society or tribe is passing so many more of its members into the future. Are these effects strong enough to show up in some future detailed research study? Will they noticeably influence our evolution overall?

Whatever the future holds for humankind, I very much hope we are all in it together, that we all continue to remain one species with wisdom enough to preserve as many other hominids and other creatures, the Ivans and Vips of the world. I hope we keep using our big brains to understand and appreciate the extraordinary process by which we came to be (and hope to remain) the top species on the planet.

34

ASTROBIOLOGICALLY SPEAKING: IS ANYONE THERE?

I intensely remember lying on the grass in my front yard in Washington, D.C., when I was perhaps nine years old. The sky was a lovely bright blue. My father, influenced by his time staring at the sky in a POW camp during World War II, was fascinated with the stars. He had let me look through his old telescope a few times; the experience left me with a vague sense that there *had* to be other worlds going around those distant points of light. On this particular day, I had just been to the National Art Gallery and seen some paintings by Van Gogh. I was intrigued that his sky was often not blue. I turned these ideas over in my mind. I remember imagining another boy like myself, living on another world, and wondering what the sky might look like there. Would it be green? Pink?

What was once the stuff of childhood daydreams is now the next frontier in evolutionary science. Since 1995, astronomers have found nearly two thousand confirmed planets around other stars. Some of these planets are similar to Earth in size and mass. About two dozen of them orbit in the habitable zone, the distance from their stars where temperatures are potentially suitable to our kind of life. Extrapolating broadly, there may be 50 billion habitable planets in our galaxy. Nowadays, there is an entire field of science known as astrobiology—the study of life among the stars.

Asking about life elsewhere is really another way of asking about living things in general. It's equivalent to asking, "Just what does it take to be a living thing?" It forces us to reexamine all of the evolutionary ideas we've discussed so far from the bottom up. At the most basic level, life clearly needs chemicals of some sort. Any realistic living thing we envision would be made of atoms and molecules, just like you, me, birds, and trees. (Sure, there are living things made of pure energy or dark matter or other exotic things in science fiction stories, but that's not the kind of hard scientific speculation I'm talking about here.) From there, things rapidly get more ambiguous.

When I was in school, it was generally agreed that in order to live, you had to have sunlight. Everyone still agrees you need a source of energy, and sunlight is a great one, but now scientists understand it is not the only possible one. In my lifetime, we have discovered hydrothermal vent ecosystems at the bottom of the sea, where sunlight does not penetrate. These systems are powered in a way that scientists had not considered until they saw it in action. The animals down there rely on certain bacteria that help them metabolize the chemical energy in hydrogen sulfide and water, and the extraordinary amount of heat energy that streams up from the ocean floor. The bacteria in turn produce chemicals for giant tubeworms, bright white crabs, and unusual huge clams as they all make a living.

In recent years, the deep-ocean hydrothermal vents have been studied extensively. As I write, it's looking likely that the organisms there got their start on the surface. They are probably all descendants of surface creatures; they are all to the right or downstream on the Timeline of Life from other clams, crabs, and worms. If so, that finding contains an important message: Life's biochemistry is flexible enough that it can switch from one kind of energy input to another. But the process had to begin somewhere. In order for the whole metabolism of life we know to get going, the chemicals need some initial energy input. Maybe life started in a hot sea and made its way to the surface. Researchers press on.

According to one particularly compelling hypothesis, the energy input in question was electromagnetic radiation in the ultraviolet range. For you Latin buffs, *ultra* means "above" in Latin. So, ultraviolet is at an energy level above the violet or purple that's visible to our eyes. The ultraviolet hypothesis points to at least two ways to explore the origin of life. First, we could check our climate computer models, and especially our computer models of how stars evolve, and see if it is reasonable that Earth was well irradiated with ultraviolet light from the Sun in life's earliest days, say 3.5 billion years ago. Second, we can study whether there is something in the genes or DNA of the ocean vent creatures that gives us a clue as to their origin and the need for their ancestors to have used ultraviolet light for their early metabolism. If this proves to be true, it would mean that living things there now started on the surface and made their ways to the ocean depths. It's a fascinating bit of research that could inform the way we look for life on other worlds, which may lead to that discovery of life elsewhere.

In coming years, we can also expect more experiments that attempt to study the origin of life directly by re-creating nature's experiment and synthesizing self-replicating molecules in the laboratory. It will then be a reasonable question: Is it alive?

Whether we are seeking out new forms of life in nature or trying to create them ourselves, here's something they almost certainly will need: liquid. Life requires some solvent to transport molecules from one place to another as it extracts energy from them, or uses them to make other chemicals move around. Astrobiologists have studied all kinds of possible life-sustaining liquids: ammonia, chlorine, liquid methane, alcohols of different varieties, and so on. They've explored their properties at different temperatures and pressures. In the end, they keep coming back to water. It is just a very effective, very versatile solvent. It also has the advantage of being very abundant.

Our solar system is loaded with water, if you know where to look. Jupiter's moon Europa has a saltwater ocean under its enormous shell of water ice. Asteroids are often made up of a great deal of ice. The giant asteroid Ceres—so big that it has been reclassified as a dwarf planet (lame expression, oh well)—seems to have a surface made of wet clay. We'll know soon enough, since the Dawn spacecraft is headed there in 2015. Pluto and its moon Charon, along with the smaller satellites Nix, Hydra, Kerberos, and Styx, are undoubtedly rich in ice. Big news coming here, too, as the New Horizons probe will be flying past the Pluto system on July 14, 2015. The whole swarm of worlds beyond Neptune, known as the Kuiper Belt, is probably full of frozen water. By the way, instead of thinking of Pluto as the last of the traditional planets, I like to think of it as being the first of a new class of objects called the *Plutoids*. Water even shows up in the most unlikely places. It's present in frigid, shadowed craters at the north pole of the otherwise searing-hot planet Mercury, and forms a frost in the polar craters of the Moon.

In any consideration of life on other worlds, a wonderful evolutionary question emerges right away. How different would any type of alien life really be? You have to figure that it would not have DNA. Or, would it? It would not have cell membranes and organelles that metabolized chemical energy. Or, would it? They, if there are any of them, would not have five fingers and toes on each of four appendages. Or, would they/it? They/it would not have a water-based brain closely connected to chemical, stereo-aural, multi-channel-tactile, and stereo-optical sensors (taste, smell, hearing, touch, and sight). Or, would they/it, etc.? Would the process of evolution converge on common designs and problem solving for the contingencies of life, or would everything be different?

If there is life elsewhere in the solar system it might actually have a lot in common with life on Earth. Think about the enormous asteroid that struck Earth 66 million years ago, when the ancient dinosaurs vanished. As it kicked up a huge cloud of rock and dust, some of that material escaped Earth entirely and began migrating through the solar system. The same thing happens on other planets. Over billions of years, the planets exchange quite a bit of material. This is not speculation; this is fact. Planetary scientists have found pieces of Mars and the Moon here on Earth, and may have identified fragments of Venus and Mercury as well. Could life have made the journey from Earth to Mars, or vice versa? This idea has come to be called transpermia, sending life across interplanetary or even interstellar space.

It's also possible there is a completely different type of life out there on one of these other worlds in our solar system. Then we could study it and gain great insight into nature and the process of evolution. Along the way, such a discovery would lead to profound changes in our beliefs as citizens of planet Earth. But we'd have to send spacecraft and researchers way out there to conduct an investigation, and we'd have to be exceedingly careful about avoiding contamination in either direction. It's a delicate business this dealing with aliens.

Scientists and the lay public alike are taking these ideas increasingly seriously. After they landed on Mars in 2004, NASA's *Spirit* and *Opportunity* rovers sent back conclusive proof that water once flowed almost everywhere on that world. At that point, the famous gambling houses in Britain such as Ladbrokes and the William Hill Company stopped taking bets on whether or not life will be found on Mars. In

2004, the betting closed at 16-to-1, down from 1,000-to-1 forty years ago. Ladbrokes is the same gambling house that paid out £10,000 to a gentleman who successfully wagered just £10 that humans would land on the Moon before the end of the decade of the 1960s. This sort of wagering is more than just human entertainment. It shows the kind of support a government or commercial space company can expect from the public in the search for life elsewhere. As the CEO of the Planetary Society, I hope that excitement persists, and inspires the kinds of missions needed to get concrete answers.

At this point, I hope you're asking yourself, "Where is the most logical place to look for alien life?" As Earthlings, we cannot help but look at Venus and Mars. These two worlds closely resemble our own, astronomically speaking. They are similar distances from the Sun. Earth is about 13,000 km in diameter. Venus is about 12,000 km. Mars is about 7,000. Neither neighboring world is quite as big as ours, but for an astronomer, it's the same order of magnitude—rounding to one significant digit, they're all three about 10,000 km across. Earth turns in 24 Earth hours. Venus takes 243 days, but Mars spins once around in 24 hours 40 minutes; a Mars day is virtually the same as Earth's.

We are getting to know Mars quite well. As of this writing, the *Opportunity* rover is still roving. It was designed to run ninety Martian days—just over three months—but it is still running, ten years later. That's like buying a car with a three-year warranty and finding it running 120 years down the road, with no maintenance, no oil changes, no tire rotations, and without new brake pads. It is an amazing example of your tax dollars at work. Meanwhile, the bigger, newer *Curiosity* rover is hard at work exploring another part of the planet.

Venus is a whole other ball of rock. It looks great from here. When I was a kid, science fiction movies filled with lovely Venusian women were a staple; earlier, some scientists seriously imagined Venus as a steamy jungle planet populated with dinosaurlike creatures. Upon closer inspection, though, Venus proves to be utterly inhospitable. Its surface is hot, astonishingly hot, around 460° C (840° F). That's intense enough to melt lead. Your fishing weights would just turn to pools of shiny metal. Venus is kept that way by a thick, dense atmosphere that's full of carbon dioxide. It's the greenhouse effect gone wild—runaway, as it is oft described. In fact, the models of climate change here on Earth were developed in part by scientists, James Hansen especially, who were studying the atmosphere of Venus. They observed that visible light passes the atmosphere, hits the surface, and then is reradiated as heat that is then trapped by carbon dioxide. This process has a big influence on whether or not a planet is habitable.

The carbon dioxide in the Venusian atmosphere was cooked out of carbon-bearing chemicals, the carbonates, in the Venusian rocks. It is so hot there that the oxygen (O) broke away and bonded with carbon (C) in the air. Any water that was once on Venus has been bonded with sulfur to make sulfuric acid (H_2SO_4), giving rise to sulfuric acid clouds. On Venus, it rains acid. But the acid rain never hits Venusian ground, because the heat is so intense that the raindrops evaporate before they can make it to the surface. Venus is a good approximation of hell. A few Soviet space probes managed to land there and look around, but even the hardiest lasted just a couple of hours. This is not a promising place to look for life, though some researchers have suggested that sulfur-loving microbes could possibly survive in its clouds.

Mars has opposite problems. The air is thin, and the surface is very cold. We can observe the ice caps of Mars with telescopes from here on Earth. There is quite a bit of regular water ice, but the enormous white arctic and Antarctic ice cap features on Mars are mostly frozen carbon dioxide (dry ice). That makes the polar places at least –130° C (–270° F). The water vapor and carbon dioxide that condense to form the ice caps are part of the Martian atmosphere. By an Earthling's standards it's not much of an atmosphere, barely 0.7 percent of the surface pressure of Earth, but it's enough to create winds and interesting weather. The *Opportunity* rover drivers from time to time direct the spacecraft to drive uphill to higher ground, where the winds and electrostatic conditions are favorable for blowing the dust off the vehicle. It's an unusual business, but one that has made remarkable discoveries.

The ongoing explorations of Mars show that the planet was once covered with lakes, streams, and expansive seas. The *Curiosity* rover practically landed in what is obviously a dry riverbed. One cannot help but wonder with all that water on Mars over three billion years ago, were there living things there? Could Martian microbes live on today, underground where they are protected from climate extremes and cosmic rays? On the open ground that our rovers are able to get to, we observe evidence of water, but nothing looks to be living there today. Keep in mind, though, that our technology is limited. The limitations can be expressed in dollars for planetary science. With our current technology and investment, we can land our exploratory spacecraft only on open areas on the Martian surface. We can't narrow the landing area enough yet for a precise touchdown point. This is a real constraint in the search for life, or evidence of life. Imagine you were to explore Earth looking for life, but your technology constrained you to land on the Great Salt Lake flats, or in the Sahara Desert. You might not see much in the way of life until you had driven hundreds of kilometers in the right direction.

The Phoenix Lander alighted upon the Martian surface in 2008. Its findings added another intriguing twist in the search for life on Mars. Phoenix landed atop a thin layer of sand or soil at the north pole. Right below the surface, just a few centimeters down, there is an enormous sheet of ice, water ice. It's apparently just below the surface for many kilometers in every direction. What if there is something living in that ice that might be akin to the half dozen or so genera of bacteria that live below the ice on our own planet? As we have seen on Earth, life is extremely tenacious once it gets started. If early Mars was sufficiently hospitable, maybe the process started there billions of years ago and never ended.

As the CEO of the Planetary Society, I often advocate for a big investment in the search for life on Mars. Suppose we built a spacecraft that could land near a valley, gulley, or gulch near the equator of Mars, a place where it might get just above the freezing point of water on a sunny Martian summer day. Then suppose we had a rover that could detach from the main spacecraft, drive over to the edge or rim of the gulley, then descend on a tether, lowering itself, like a rappelling rock climber, to an icy outcrop or exposed stratum. In the midmorning as the Sun shone directly on that ice, instruments onboard our tethered rappelling robot would look very, very closely. What if they found something still alive there? What if there are indeed Martian microbes still eking out a living way out there in the icy cold of our next neighbor?

Answering these kinds of questions is not terribly expensive, in the bigger scheme of things. Right now, the U.S. investment in planetary science is less than $1.5 billion a year. Put another way, it is less than 0.05 percent of the federal budget. That includes all the missions: Mars, Mercury, Jupiter, Saturn, and the New Horizons mission currently en route to distant Pluto. What if we upped that a billion and found life on Mars? It would be an extraordinary investment, costing barely the equivalent of an extra cup of coffee per taxpayer. With a president, a Congress, and a NASA administrator focused on such a thing, we could change the course of human history.

The same could be said for a trip to Europa, one of the four large satellites of Jupiter. Europa is 3,100 kilometers in diameter, just a bit smaller than Earth's moon, but it is an entirely different type of world.

In 2011, data from the *Galileo* spacecraft were analyzed carefully. It is clear now that there is a salty ocean of water under Europa's heavily cracked surface shell of ice. The ocean was discovered using magnetometer data; it's a sensitive electronic compass. Salt water conducts electricity, which in turn affects the magnetic field around Europa. The water has not frozen solid, because the orbital motion of Europa in Jupiter's powerful gravitational field makes the whole world squeeze and unsqueeze with each orbit. Europa maintains its liquid ocean with heat generated by mechanical distortion. It's just like the warmth you feel if you stretch an uninflated rubber balloon a few dozen times and then touch it to your lips. Try it!

Ever since that discovery, scientists and engineers have been discussing how to explore that ocean under the ice. If there really is liquid water, and it really has been kept warm enough to remain a liquid these last four and half billion years, perhaps there is something living there. Plans have been drawn up to build a spacecraft that would land on the surface of Europa. It would then deploy a mechanical or thermal drill, a penetrator tough enough and perhaps hot enough to work through up to fifty kilometers of ice. It would be tethered to the landing craft up above. It would carry instruments that would look for, what we imagine would be, signs of Europan life. That would be a thrilling mission, but a very costly one, certainly many billions of dollars. It would be enormously technically challenging. And of enormous importance, it would have to be careful not to violate science fiction's "prime directive." To wit, we must not screw up the Europan ecosystems, if there are such things, by contaminating it with Earth microbes that hitched along for the ride.

In 2013, we discovered something exciting, something that might greatly simplify the search for Europan life. Astronomers aimed the Hubble Space Telescope at Europa and discovered plumes of water, seawater, spewing right out into space through fissures in the ice. If there are microbes, or even maybe centimeter-sized living things in the Europan ocean, they're being squirted right along with water into the blackness. A spacecraft could be designed to fly through the plume of water, capturing plenty enough of it to do microscopic and chemical assays of whatever might be living in that water. Such a mission can be flown for a small fraction of the cost of a lander with a drill and all the trouble we would have to prevent our microbes from contaminating theirs (if there are any of "them"). The proposed mission is called the *Europa Clipper.*

A similar challenge awaits at Enceladus, one of Saturn's moons. It is much smaller than Europa, just five hundred kilometers wide, but it, too, has a (small) ocean of water buried beneath a thick ice pack, and has even bigger jets that erupt from its south pole. Here is another place to look for life using a *Clipper*-style mission.

I don't know about you, but I find it easy to imagine some sort of deep space–worthy transparent plate mounted roughly perpendicular to the spacecraft's direction of flight. As the *Clipper* flew through plumes of water any living thing might end up there like insects on a windshield. (Not perfect, but it may be the best we can do; orbiting takes velocity.) Then with a microscope camera with an appropriate light source that could be trained on the plate, Earthlings would get a glimpse of what might be a living thing from an alien world. A more elaborate version could even collect a sample of Europan (or Enceladean) ice and bring it back to Earth for more detailed study—taking a lot of care not to contaminate things either way, of course. Talk about a tax investment value: This would be a groundbreaking experiment unlike any performed ever before in the history of the world.

If we go to Europa or Enceladus—or search more aggressively on Mars—we will run into a higher-level question about possible alien life: Will we be able to recognize it if we see it? The answer requires going back to what we know about the fundamentals of life, beyond its need for energy and its probable fondness for water. How would it regulate the chemical reactions that are needed for chemicals to make copies of themselves? One way that we know works is to use the chemical properties of chemicals already dissolved in water to provide the energy to move things around. But a living thing needs some way to keep different chemicals separated. Otherwise, the whole insides of a living thing would probably just mix up and come to a stop. So, we figure that it will need a container or membrane. It needs a wall to establish what's inside, and what constitutes its surroundings. In short, it might be very different on the inside, but from the outside it is likely to look like the bacteria and single-cell organisms we know so well, at least by one logical line of reasoning.

Living things that can form membranes would probably have a big advantage over other molecules that can't form them. Membranes enable living things to use the attraction and repulsion of the electrons on the outside of atoms to drive or pull molecules around. Try the osmosis experiment I mentioned back in chapter 12. It applies equally well to chemical systems on other worlds.

Searching for evidence of water on other worlds is very straightforward. Searching for membranes is a much more complicated business. Let's go back to the case of Europa. Suppose we build the *Europa Clipper*, it flies through the geysers, and it brings samples back home. Even then, how would we know? How would we find cell membranes amidst the spray? One idea is to look for the kinds of atoms that we find in the membranes of cells on Earth. In terrestrial life, the characteristic elements in a membrane are carbon, nitrogen, potassium, and sodium. There's a place to start . . . unless Europan life came up with a totally different way to make a membrane. Then how would you find it? If you like this kind of thinking, consider becoming an astrobiologist.

The problem is somewhat easier for Mars, because it is much closer and because there is a more straightforward, Earthlike rocky surface to explore. In the coming years, we will continue to search for life there. If we could get the right instruments to some super salty slushy outcropping, we might find evidence of fossilized Martian microbes. We might even find something still alive, and observe it directly through a microscope.

Just think what it would mean if one of those distant bodies is spewing some heretofore-unknown type of life into deep space, or if life awaits sheltered under a rock on Mars. What if that life is like us? What if it's totally different? Whatever the answer, the discovery would change the way we all think about what it means to be a living thing. It will tell us about the different ways in which life can arise and evolve. It would give us, for the first time in history, concrete proof that we are not alone in the universe.

In my wilder moments of speculation, I like to go even further. I imagine that there might be something swimming in the sea under the ice of Europa—not just microbes, but big, complex organisms. If there really is a whole ecosystem in the Europan ocean, and it's been there long enough to have established something reminiscent of our Earthling multicellular sea creatures, I assume they'll be fish-like in shape. It seems to me it's even reasonable to expect any organism—fish, fowl, or fruit fly—to have its sensory organs concentrated in something like a head and its locomotion appendages somehow wired to its head, and so on.

In other words, I wouldn't be surprised to find alien creatures with body plans not too different from ours. I have no idea if I, or anyone, will get to see such a thing. But it sure would be a powerful test of our ideas about convergent evolution. What would we learn from an alien? It could be astonishing. We'd quickly find out if life necessarily needs a genetic code, a cell membrane, similar kinds of appendages, and familiar sorts of sensory organs. Nature may have possibilities we haven't even imagined. Or maybe life, like impact craters or volcanoes, tends to look pretty much the same everywhere it occurs.

These ideas would merely be arcane musings were it not for the fascination we all feel about understanding our origin. There is no more dramatic way to test, and extend, the limits of what we have learned about evolution than by searching for evidence that it has occurred on other worlds as well. The answer will drive new technologies. It will inspire future generations of scientists. And it may revolutionize both our practical and our philosophical understandings of what it means to be human.

35

THE SPARKS THAT STARTED IT ALL

In his expansive discussion of evolution in *On the Origin of Species*, Charles Darwin assiduously avoided contentious speculation about how the whole story began. His commentary is restricted to a single sentence near the end of the book's last chapter: "Therefore I should infer from analogy that probably all the organic beings which have ever lived on this earth have descended from some one primordial form, into which life was first breathed." But the question is irresistible. Where did we come from—what was the spark that lit life's fire? These days, many scientists are venturing where Darwin could not dare. Let's join them and go back to the beginning to talk about . . . the beginning.

Asking the big question sounds an awful lot like asking, "Is there a god who runs the show?" There is an essential difference, however. Every other aspect of life that was once attributed to divine intent is now elegantly and completely explained in the context of evolutionary science. For me, there is no reason to think that the origin of life is any different. I am open-minded, and have no problem with most religions, but religious explanations are unsatisfactory. They don't take me anywhere; you either believe them or you don't, and that's that. Scientific theories of the origin of life are open to questions, to tests, to revisions, to replacement with new and more insightful theories. One path leads to a dead halt. The other leads to thrilling, limitless forward motion.

When I was in engineering school at Cornell University, I moseyed over to the Space Sciences building now and then. John Olsen (Jolse) was and is a good friend, a fellow serious bicyclist and very much into astrophysics and he always nudged me to attend symposia and small graduate-level talks, etc., up there in Space Sciences. As John often pointed out, it's just wild, the things they talked about: black holes, the center of the universe (or lack thereof), energy production, and the synthesis of new elements in stars. These meetings included Carl Sagan, Kip Thorne, and Hans Bethe. What a time to be hanging out in that not especially good-looking cinder block building. On one of my wanderings, I ended up in a lab, which I'm pretty sure was on the third floor, and there before me was a fabulous tangle of glassware and tubing connecting large metal bottles of various gasses to a very large central spherical flask. It was a version of the Miller-Urey experiment, eagerly sparking away.

Hold on; let me back up for a moment. The idea of these setups, which were first proposed, designed, and run in the 1950s by chemists Stanley Miller and Harold Urey, was to simulate the conditions on Earth in primordial times, three or four billion years ago, when life appeared for the first time. The experiments were intended to see if they could make something come alive using nothing but nonliving chemicals.

Do you know what emerged? Not life, but something fascinating all the same. The chemicals gave rise to significant chemical compounds: a handful of amino acids, essential components in the chemistry of life. Amino acids are molecules that hook together to form the proteins that run almost every aspect of biology. They are the building blocks of living things. The details are fascinating, but as a general description, acids are chemicals that can yield or "donate" a proton to another atom or molecule. An acid can be weird and deadly, or it can be mild or gentle like salad dressing.

When it comes to amino acids, they all have a single carbon atom in the middle and a carbon-oxygen-oxygen chain on one side. The remarkable property of carbon is that it carries four places for other chemicals to attach, four "bonding sites" as they're called. In amino acids, one of the sites is a chain of atoms: carbon-oxygen-oxygen. On its own, we call it carboxylic acid. When it's connected to another molecule we refer to it as a carboxyl group. In amino acids, one of the central carbon's sites is for the carboxyl group, and the other three sites carry various other configurations of carbon, sulfur, nitrogen, and especially hydrogen.

I mention all this detail, because it's astonishing to realize that that's it. There are only twenty or so naturally occurring amino acids. (There's some debate about exactly how many are used by living things, but at most it may be a few more than twenty.) They all come from combinations of just a few different types of atoms. Amino acids form the so-called peptide molecules that link together to create the polypeptides that form proteins. Proteins, in turn, do much of the work in cells. They build structure; they run metabolism; they regulate responses, the whole shebang. And these amino acids were produced in the several versions of the Miller-Urey experiment. Actually, they produced the natural amino acids, along with a few additional but nonnatural, chemically logical configurations of other amino acids. It's astonishing. Just five chemical elements, and look at all the living things they create!

To create these compounds, Miller and Urey had to infer the composition of the primordial Earth's atmosphere and the primordial Earth's ocean. With the advantage of a few decades of additional knowledge, biologists today generally believe that they played it more conservatively than perhaps they needed to. They loaded the big glass flask with some natural gas or methane, some water vapor, and some ammonia. If you're scoring along with us, those are CH_4, H_2O, and NH_3 in chemical notation. They gave it a spark, which they reasonably figured would have happened during primordial thunderstorms. One key missing element: They didn't have a source of sulfur.

It's a safe bet that there were a lot of volcanoes erupting all over the place on the young Earth, spewing out the rotten-egg smelling hydrogen sulfide (H_2S). It's a deadly poison to animals like us now, but in those long-ago times it may have led to an important amino acid called cysteine. Hydrogen sulfide might also have been useful to early life as a primary energy source. Even today there are whole ecosystems in deep-ocean hot water vents that run on hydrogen sulfide. One organism's trash is another organism's treasure, as I like to say.

It is interesting for me to note that creationists generally dismiss Miller-Urey style experiments, saying that the idea of life arising from chemicals that are not imbued with some divine power is preposterous. One reason creationists dismiss these results is that they say the quantities of amino acids produced are so small as to be insignificant. That is just plain wrong. Any quantity of the basic molecules of life is infinitely more than zero—infinitely more. The origin of life just requires some raw material that could allow the spark of life to emerge. Evolution is a powerful amplifier. Once a self-replicating system is established, it has a chance to scour the environment for resources and systematically make more copies of itself. Even 0.79 percent of a 10-liter volume—the concentration of amino acids in the Miller-Urey experiment—might very well amount to enough to start life on its way in the uncontested environment of the primordial Earth.

From there, this idea that creationists call molecules-to-man is quite reasonable, because a lot can happen in 3.5 billion years. The process is often called abiogenesis (life from not life), and it is still the leading theory of how biology got started on Earth. Abiogenesis may be what led to you and me. An aside: At one time abiogenesis was used to describe another pseudoscientific presumed phenomenon called "spontaneous generation." Barnacles, for example, seem to grow out of nowhere or nothing. It was just that the people making those observations didn't use magnifying lenses. Barnacle polyps are tiny, yet they grow into shelled creatures that are quite tough. They were in the seawater all along. That adult barnacles seem to generate themselves spontaneously is just a product of not looking closely enough.

It's important to keep in mind the enormous amount of space and time that life had to work with. Experiments like the one crafted by Miller and Urey used a system the size of a laboratory flask; the surface of Earth is about a trillion times greater. The experiment ran for two weeks, whereas life on Earth had something close to a billion years. Furthermore, I suspect that these experiments were missing a key ingredient, a key type of electromagnetic, or electrical, or chemical energy, and if we figured it out, we could create molecules that made replicas, even crude replicas, of themselves. There is just so much we don't know about what conditions were like on the primordial Earth. And by the way, what's to keep real abiogenesis from happening here on Earth right now? Something to think about, yes?

Taking a very different approach, Craig Venter—the scientist who first sequenced a complete human genome—asserts that he has already created artificial life. He and his team separated or captured a bacterium and determined its gene sequence. They then created their own synthetic or novel DNA and inserted that into the bacterium, converting it to a new artificial strain. It reproduced like crazy, creating billions of a new human-made or artificial species of bacteria. Unlike Miller and Urey, Venter is not aiming to create life purely from raw chemicals, at least not so far. Still, the result—it's wild.

Venter's immediate plans are fairly pragmatic: to create artificial bacteria that can produce renewable fuels and new medicines. Just creating a chunk of synthetic DNA took an enormous amount of trouble. But of course, the longest journey starts with but a single step. The same researchers produced an artificial virus seven years earlier. They implanted a genome of their own design into a living cell (a bacterium), and sure enough the cell was directed to produce copies of the implanted gene sequence. It was an artificial virus. It could not live on its own, but it could induce a living cell to make copies of the artificial virus, if I may, just like a natural virus. By the way, Venter Institute is very carefully watched for sound ethics and microbial safety practices.

Research has come a long way since the Miller-Urey experiments. Investigators have studied the growth of complex carbon-based molecules in ice. They've studied the chemistry of clays that have strong alkaline properties, which might induce chemical reactions that would give a nascent molecule enough of a jolt to start replicating. They've looked at possible early life based on the molecule called ribonucleic acid (RNA), the simpler cousin of DNA.

Very recently, research led by MIT physicist Jeremy England suggests that life may happen automatically, a result of physics, specifically thermodynamics. Professor England argues that molecules assemble themselves in the most energy-efficient way they can find. Molecules may be driven to seek thermodynamic equilibrium, and that may lead to life. The idea is as compelling as it is wild.

Researchers have even looked hard at the possibility of life starting on another planet, Mars being the logical one, and finding its way here through interplanetary space. And besides, would we even know a nonliving thing from an extraordinarily slow-growing living thing, even if it were to literally bite us in the leg, albeit on the microscopic level? Yes, it probably would have a membrane, but unfortunately a lot of small nonliving things are also round or rod-shaped, just like bacteria.

For the first 3 billion years of life on Earth, as we know, the progression from self-replicating molecules to the Cambrian Period, when fossils got big enough to see, is still a bit unclear. What if we found life like Earth's own primitive life, microscopic and soft? Would we recognize it?

I remember very well my older brother coming home from school and posing the question: Are viruses alive? Are they living things? You can leave them in a jar for years, and they'll pop right back out and start doing what they used to do. In the presence of other organisms, they reproduce like mad. They mutate. They interact with other cells. But when they are by themselves, they exist in a kind of stasis. My brother and I had a protracted discussion and came to the conclusion that the answer is clearly: Maybe . . .

Looking at them from where we are on the Timeline of Life, it's clear that viruses do reproduce. They induce cells to do the reproducing for them. A cell, any cell, cannot live without its environment, something like drifting in a chemical broth or being stationed close to a blood capillary. So from this standpoint, a virus is a living thing that, instead of having chemical nutrients around, it just has to have another living thing around. It is obliged to live in and among cells—other living things. So, we call it an "obligate parasite."

But what does it take for us to consider an organized bunch of chemicals to be a living thing? Astrobiology research helps guide us here. Generally, we want it to have a membrane. We expect it to reproduce, and we want it to maintain a steady chemical balance or steady state within itself. In biology, we say a living thing by definition can maintain equilibrium, the official word being *homeostasis* ("stays the same").

When it comes to bacteria, no question. There they are with membranes and walls, they keep their metabolisms going in different surrounding conditions, and they certainly reproduce. When it comes to viruses, it depends how you look at it. Viruses do not maintain homeostasis. From the virus's point of view, why bother? If your molecules can all hold their arrangement in what might be considered hostile environments, why bother doing anything else? Why complicate things, when your systems work fine as they are? To my way of thinking, we just wouldn't have viruses, or the whole domain Vira, if we did not have the domain Bacteria. So they belong on any depiction of categories of living things on Earth.

In fact, viruses could be the key to this whole conversation, because they may show how life emerged in the first place. Could viral molecular structures have existed before self-replicating bacterium-like molecules came to be? Or, are viruses a side product, an unintended consequence of the natural processes that brought about self-replicating molecular arrangements in the first place? Viruses are simpler than bacteria, which would seem to make them good candidates for arriving earlier.

There are other aspects that make them not look like a very good model for the first living thing, however. For one thing, unlike bacteria (and you and me), viruses do not seem to have a common ancestor. In the strange world of viruses, there is nothing akin to the way that all other living things use DNA to reproduce. Each virus attacks or latches onto specific, sometimes surprisingly specific, living cells. When a virus attacks, it does not exchange genes, or transfer genes with or from the cell it's attacking. A virus's genes go into a cell, never the other way around. So, in these important traditional senses, a virus is not alive. Viruses seem to have come into existence after self-replicating, stasis-maintaining organisms (the things we consider definitively "living") came to be. But let's face it: Viruses are a very significant part of our world. Without viruses, every living thing would be different.

Viruses are somehow part of the spectrum of life, and they are somehow related to the question of how it all began. We are still sorting out the thick and tangled branches of the Tree of Life; only within the past decade have biologists identified the giant viruses that help build the case that they qualify as a fourth domain of life. The harder we push on the search for the origin of life, the more surprises we find. These discoveries keep me wondering: "Mother Nature, what else haven't you told me?"

36

A SECOND GENESIS—OF LIFE?

I was a child of the space age; I crouched in front of a black-and-white vacuum-tube television in 1969, when humans first set foot on the Moon. When I think about an alien form of life, with all that it would tell us about the process of evolution, I naturally picture something on another world. But a few researchers, including Paul Davies at Arizona State University, offer a drastically different perspective. They suggest that aliens—microbial aliens, that is—might be right here on Earth, right now. If these researchers are right, we've been missing out on one of the most powerful ways to understand the origin of life. There might be a whole other Tree of Life all around us hiding in plain sight.

The possibility of a second type of life getting a foothold here on Earth is called the second genesis. If we could find a second kind of living thing, it would change our world. Right now everything we know comes from an example of one. If we could compare two totally different types of organisms, we could begin to find out experimentally whether there are alternatives to DNA, other kinds of metabolisms, whole other life chemistries. The practical implications are as mind-boggling as the scientific ones. And I wonder: What would the creationists say?

I can almost hear your objections. How could there be unknown organisms on Earth, considering how thoroughly biologists have studied all the microbes around us? Well, think about the challenges of recognizing alien microbes. The same is true right here at home. There are two basic ways to find a new microbe: You can grow it in a petri dish, or you can sequence its DNA. If it doesn't eat the same things that other microbes eat, and if it doesn't have the standard kind of DNA (or doesn't use DNA at all), it would basically slip through the cracks of modern biology.

A huge lesson from the past few decades is that Earth is full of surprising and unexpected organisms. When we examine oil well tailings (the rocks and soil brought up by the drill bits), we find entire ecosystems of slow-growing bacteria. These organisms date back millions of years and they've never seen the light of day. They still have DNA. We have a common ancestor. But they are drastically unlike anything that lives on the surface.

One cannot help but wonder what else is down there. Is there a system of living things that we know nothing about? In an area referred to as the Illinois Basin, there are enormous coal mines extending kilometers in several directions underground. When you look around down there you cannot help but notice the scale trees, an extinct fernlike species that lived in primordial swamps about 300 million years ago and later turned to coal. There were probably whole microbial ecosystems in those areas that we know nothing about. These would be types of living things that we don't recognize today. They may be so different that we wouldn't know them even if they were staring right back at us up through our microscope lenses.

There may be countless places to look. Most of Earth is covered by the ocean. There is far more wet land underneath the sea surface than there is dry land above. There are billions of tons of silt and detritus down there. Perhaps a different style of life lives in the ooze. Under the Antarctic ice there is a freshwater lake that had been untouched by surface creatures for millions and millions of years. Maybe a spark was lit down in the lightless cold that still drives an alien life form even today. There are extraordinarily remote places high in the Himalayan and Andes mountain ranges. Is there a niche where an alien lifestyle is being lived? Maybe we've had samples in our labs and just overlooked that life, because it is alien in so many ways. Any of the organisms in these examples would be from our Earth but completely alien to us. Imagine it: an entire new domain of living things—no, a whole separate Tree of Life, the result of a whole separate, second genesis.

It's an exploration we must pursue. It could lead to a new branch of life science. Whatever the second genesis life forms are like, you can bet they have followed the same principles that our ancestors did. The other life may be on a different course, but I'm sure we play by the same evolutionary rules.

You might call this the outer limit of evolutionary thinking. Darwin looked at organisms as they are; we are speculating about organisms that might be. When he drew his first Tree of Life, Darwin was looking at the scheme that connected all living things; we are looking for organisms that stand stunningly apart. Yet, none of this would be possible without Darwin's discoveries and without the inquisitive spirit of evolutionary thinking that he embodied.

37

LIFE'S COSMIC IMPERATIVE

At the end of the journey of this book, I find myself thinking about the end journey of evolution itself. Not just human evolution, but the process as a whole. I have long wondered whether there is some kind of cosmic imperative for life to spread from world to world. On Earth we see that organisms colonize every possible niche. We see that living things move into new environments and beget new species. Is life destined to colonize the solar system, the Milky Way, and ultimately to spread out through the whole universe? Darwin described the "Struggle for Life" that drives evolution on this planet; perhaps in the future that same struggle will expand to a cosmic scale. Are you and I genetically driven to build starships and wander the universe? Wow, it knocks me back a little just writing it.

You may have picked up hints of this idea in the past few chapters, when I talked about the ways in which life might arise on different worlds or get randomly spread from planet to planet. But it's not hard to imagine that someday humans might start spreading life deliberately. As we explore the planets, asteroids, and comets of the solar system we already have to be extremely careful not to bring microbial life with us. What if our truly space-faring descendants choose to create colonies, and bring other living things with them intentionally? In understanding the process that led from a simple cell to us, we are beginning to sketch out a grand evolutionary process. Going from world to world, or even star to star, is entirely consistent with the ways in which life evolves to fill every possible environment on Earth.

Even in this far-out realm of speculation, others have already begun taking measure of the lay of the land. We could purposely leave genetically modified microbes on Mars that would somehow change the atmosphere there to have more oxygen, and thereby be more habitable. Perhaps we could do the same for something in the cloud tops of Venus. Craig Venter has raised the idea of a DNA fax that could read the genome of (possible) living things on Mars and send the data home so we could re-create them here on Earth. It's not so far-fetched that a reverse technology is possible: We could send bioreactors to other worlds, complete with instructions to assemble microbes adapted to the local environment at the other end.

Keep going, and the ideas just keep getting wilder. If we are not alone in the universe, we might *really* not be alone in the universe. What if every civilization throughout the galaxy or universe does the same thing by accident—or on purpose? What if we are the descendants of some such galactic seeding campaign, Johnny Appleseed's trek writ galactically large?

The only way to get the answers is to keep looking at living things and learning more about the process by which we all came to be. Evolution happens here no matter how we all got started. But now we can start to ask meaningfully about origins and destinies as well.

We will go to thrilling places—unimaginable places—if only we keep our minds open to new ideas, our faculties keen to significant pieces of evidence, our youthful curiosities forever engaged.

Where did we come from? Are we alone? Search on!

ACKNOWLEDGMENTS

This book would not have been possible without my parents Ned and Jacquie, in whose home science was celebrated and academic achievement was expected. I wouldn't have ever come to be if it weren't for my sister, Susan, who made me do my homework, and my brother, Darby, the funniest and most thoughtful man I know. Many thanks to my academic colleagues, Don Prothero, Michael Shermer, and Eugenie C. Scott, without whose guidance that debate in Kentucky and a great deal of this book would have been a mess. Nina Jablonski took a great deal of time to help me and our excellent fact checker Kate Baggaley make the skin color argument accurate and compelling; thank you, thank you.

It was my editor, Corey Powell, who, along with his quick-witted journalistic guidance, convinced me to "bottle" my voice in compact chapters. I'm grateful for all his help. All this work would not have been possible without my trusted agents, Betsy Berg and Nick Pampenella, along with their assistant, Ariella Mastroianni; thanks to all. Jennifer Weis at St. Martin's Press directed me to take on this project;

thank you indeed. And everyone knows, I owe tremendous thanks to my trusted attorney, Andy Salter, and my amazing assistant, Christine Sposari.

Finally, the Stars of the Show:

I am so very thankful for all my remarkable teachers, but most especially Mrs. McGonagle, Mrs. Cochran, Mr. Lawrence, Mr. Flowers, Mr. Cross, Ms. Hrushka, Mr. Morse, Mr. Lang, and Professor Sagan. Each of these people influenced me in extraordinary ways. I would be a very different person without them; I am a very fortunate man.

BILL NYE

THE DELAWARE SHORE

Many wonderful contingencies brought me into this project. I owe my mother for making my childhood self hunt down a paleontology curator when she couldn't answer my dinosaur questions. At *Discover* magazine, Tina Wooden coached me through countless conversations with creationists; another dear *Discover* colleague, Pam Weintraub, challenged me to become a better writer and introduced me to Jennifer Weis, the devoted editor of this book. My wife, Lisa Gifford, supported me generously during the many times when I seemed to vanish (both intellectually and temporally) from my family.

And I am honored to have worked with Bill Nye, who inspired me throughout with his passionate commitment to spreading knowledge and changing the world.

COREY S. POWELL

SOMEWHERE IN BROOKLYN

INDEX